# JAGUAR
# XJ SERIES

## Other titles in the Crowood AutoClassics Series

| | |
|---|---|
| AC Cobra | Brian Laban |
| Alfa Romeo: Spider, Alfasud & Alfetta GT | David Styles |
| Aston Martin | Jonathan Wood |
| BMW M-Series | Alan Henry |
| Ferrari Dino | Anthony Curtis |
| Jaguar E-type | Jonathan Wood |
| Jensen Interceptor | John Tipler |
| Lamborghini Countach | Peter Dron |
| Lotus Elan | Mike Taylor |
| Lotus Esprit | Jeremy Walton |
| Mercedes SL Series | Brian Laban |
| MGB | Brian Laban |
| Porsche 911 | David Vivian |
| Sporting Fords: Cortina to Cosworth | Mike Taylor |
| Sprites and Midgets | Anders Ditlev Clausager |
| Triumph TRs | Graham Robson |

# JAGUAR XJ SERIES

## *The Complete Story*

## Graham Robson

First published in 1992 by
The Crowood Press Ltd
Ramsbury, Marlborough
Wiltshire SN8 2HR

**British Library Cataloguing in Publication Data**

A catalogue record for this book is available from the British
Library

ISBN    1 85223 689 2

**Picture credits**
The majority of the photographs in this book were supplied by the
Motoring Picture Library, Beaulieu. Additional illustrations were
supplied by Jaguar Cars Ltd and the author.

Typeset by Chippendale Type Ltd., Otley, West Yorkshire.
Printed and bound in Great Britain by BPCC Hazells Ltd
Member of BPCC Ltd

# Contents

|   | Jaguar XJ Evolution | 6 |
|   | Preface | 7 |
| 1 | Building up to the XJ6 | 9 |
| 2 | XJ6 – The First Five Years | 21 |
| 3 | XJ12 – The World's Most Refined Car? | 37 |
| 4 | Series II – Broadening the Range | 55 |
| 5 | XJ-S – A Continuing Story | 71 |
| 6 | Series III – Re-touching by Pininfarina | 99 |
| 7 | XJs in Motorsport | 114 |
| 8 | XJ40 – The New Generation | 126 |
| 9 | JaguarSport | 154 |
| 10 | Testers' Opinions | 161 |
| 11 | Parts and Clubs | 173 |
| 12 | What Next? | 179 |
|   | Index | 190 |

# Jaguar XJ Evolution

### Series I models – 1968 to 1973

| | |
|---|---|
| October 1968 | Original XJ6 four-door saloon goes on sale, on 9ft 0.75in (2,762mm) wheelbase. Manual or automatic transmission, 2.8-litre or 4.2-litre six-cylinder XK engines. No Daimler-badged cars at this stage. |
| October 1969 | Daimler Sovereign (equal to Jaguar XJ6) model launched. Thereafter all cars available with Jaguar or Daimler badges. |
| July 1972 | Launch of V12-engined Jaguar XJ12/Daimler Double-Six. Automatic transmission only. |
| October 1972 | Introduction of longer-wheelbase (9ft 4.75in or 2,864mm) bodyshell as option on 4.3-litre and V12-engined cars. |
| September 1973 | Series I cars superseded by Series II cars. |

### Series II models – 1973 to 1979

| | |
|---|---|
| September 1973 | Series II range takes over from Series I models. 2.8-litre model dropped from home market, export market soon followed suit. Two-door Coupé models, previewed but not available until 1975. |
| September 1974 | All four-door saloons now to be built on longer wheelbase shell. |
| May 1975 | Fuel-injection replaces carburettors on V12-engined cars. Two-door Coupés now in production. |
| June 1975 | Introduction of 3.4-litre XJ6/Daimler saloons; this engine not available in two-door Coupé style |
| November 1977 | Production of all two-door Coupé types discontinued. |
| March 1979 | Series II cars superseded by Series III cars. |

### Series III models – starting in 1979

| | |
|---|---|
| March 1979 | Introduction of Series III models, with restyled cabin (by Pininfarina). Choice of engines as for last Series II, but 205bhp/fuel injection now on 4.2-litre model. |
| Autumn 1983 | Rejig of model names, but no basic technical changes. |
| October 1986 | Series III six-cylinder cars dropped in favour of new 'XJ40' range. V12-engined cars continue into early 1990s. |

### Second-generation XJ6 – 'the XJ40' – starting in 1986

| | |
|---|---|
| October 1986 | Introduction of new range of four-door saloon XJ6 models, on 9ft 5in (2,870mm) wheelbase. Choice of 2.9-litre or 3.6-litre 'AJ6' six-cylinder engines, five-speed manual or four-speed automatic transmission. |
| October 1989 | Original 3.6-litre model dropped, in favour of 235bhp 4.0-litre version. |
| October 1990 | Original 2.9-litre model dropped, in favour of 200bhp 3.2-litre version. |

### XJ-S Coupés, Cabriolets and Convertibles

| | |
|---|---|
| September 1975 | Original V12-engined XJ-S model introduced, as coupé, with choice of manual or automatic transmission. |
| March 1979 | Manual transmission no longer available. |
| July 1981 | XJ-S HE (High Efficiency) introduced, with engine improvements, and many style/equipment changes. |
| October 1983 | New XJ-S 3.6 Cabriolet introduced, with AJ6-type six-cylinder twin OHC engine, and five-speed manual gearbox; no automatic transmission option. |
| July 1985 | New V12-engined Cabriolet added to range. |
| February 1987 | Automatic transmission available on 3.6-litre models. |
| September 1987 | XJ-S 3.6-litre Cabriolet discontinued. |
| February/April 1988 | V12-engined Cabriolet dropped, new V12-engined Convertible added to range. |
| April 1991 | Face-lifted XJ-S range introduced, comprising 4.0-litre six-cylinder Coupé, V12 Convertible and V12 Coupé. |

# Preface

It was almost inevitable that I would want to write a book about a Jaguar car one day, for this is a marque that I have admired for many years. To my eternal chagrin, I was never rich enough to buy an E-type when such cars were current models in the 1960s and 1970s, and nowadays they don't seem to fit me, somehow . . .

By almost any standards, whether in styling, roadholding, refinement, versatility and, above all, sales and longevity, the original XJ6 family, and all its offshoots, set new records for Jaguar. It was introduced when Jaguar's reputation was on the crest of a wave, it kept doggedly on when that reputation had slumped, and it soared gloriously again when the renaissance of the 1980s began. I admired it then, and I admire it now – which explains why I have tried to cram as much as possible of the 'complete story' into this book.

Like most motoring writers of my generation, of course, I was always vastly impressed by the XJ6, the V12 engine that followed it, and by the resourceful way that Jaguar made so much, and found so many derivatives, from one basic design. It was a family of cars whose fortunes I have followed closely over the years.

When I was very young, my father's unattainable 'dream car' was a rakish 3.5-litre SS-Jaguar saloon. Like many other schoolboys, I also fell for Jaguar in a big way when the motor racing successes of the 1950s began to build up. Even by that time I had decided that I wanted to work in the motor industry, so as a young and barely qualified engineer, I joined Jaguar to 'learn the trade'.

It was a period in my life that I will never forget, for I was involved, however humbly, in the design and development of cars like the Mk II and Mk X saloons, and of the E-type sports car. It was also the period in which Sir William Lyons completed the take-over of Daimler, and began laying his plans for the late 1960s.

Even though I was working at Jaguar on a generation of cars produced *before* the arrival of the XJ6, it still allowed me to observe the collective genius of Sir William, Bill Heynes, Bob Knight, Malcolm Sayer and their staff at work. With such an array of talent, which all seemed to flower under the firm way in which Sir William ran his own company, the success of the XJ6 family of cars was almost inevitable.

Because this book spans more than twenty tumultuous years, several different Jaguar chairmen, several changes of ownership, profits *and* losses, two generations of car, three types of engine, saloons and coupés, Jaguars and Daimlers, there is a complex story to tell, and I hope I have done it logically. I promise you that I have done it with affection for a series of fine designs, and I have tried to expose every feature, good or bad, which a Jaguar enthusiast should know.

I seem to have driven a lot, heard a lot, and built up strong feelings about the products. In nearly twenty-five years, the joys of driving silky V12-engined machines with astonishing performance have sometimes been tempered by the irritation of quality problems. The disappointment of seeing Jaguar merged with BMC, then nearly being sunk by British Leyland, was more than balanced by the euphoria of seeing the company privatized and successful again in the 1980s.

*1991 XJ-S Convertible*

Through all this, however, the XJ6 range survived, prospered, and made its mark on the worldwide motoring scene. There is more – much more – to add in the 1990s.

Memory and good archives help a lot, but I couldn't possibly have completed my researches without the help of many kind people. In particular, David Boole and his colleagues at Jaguar – including Joe Greenwell, Colin Cook and Val West – opened many doors for me. Joe, in particular, made it easy for me to visit Jaguar's factories, and the Whitley Technical Centre, to get a feel for the 'Jaguar way' of producing cars in the 1990s.

At JaguarSport, Bill Donnelly (Sales and Marketing Manager) and Mike Moreton (Project Manager, XJ220) gave a lot of their time, and experience.

Over the years, I have had many contacts with important Jaguar 'characters', of whom Walter Hassan, Harry Mundy and Bob Knight all gave me a great deal of help.

Anders Clausager, the distinguished archivist of the British Motor Industry Heritage Trust, was an informative, reassuring and soothing source when it came to the time to dig for statistics.

And last, but by no means least, I couldn't possibly have had the insight to write this book if I had never worked at Jaguar Cars at all. For making that possible, therefore, I want to thank the late William (Bill) Heynes for taking me on, as an engineer, in the 1950s.

Graham Robson 1992

# 1 Building up to the XJ6

Nothing is more important to the launch of a new car than its ancestry. If the new model has no heritage, no tradition, no breeding or – in horse-racing terms – no 'form' – it has an uphill struggle to be accepted. If it comes from a well-known stable, on the other hand, it is likely to be given a good reception even before the critics get behind the wheel. This, above all, explains why the Jaguar XJ6 was so enthusiastically welcomed in 1968, for it was the latest, the most attractive, and the most modern of a whole family of excitingly engineered cars.

Although the XJ6 itself was only conceived in the mid-1960s, the story behind the car stretches back into the 1920s, to Lancashire, where a young man's first connection with the motor industry was a small company building cars and trucks in Manchester. That man was William Lyons – Sir William, as he became in later years.

Until the 1970s, the Jaguar story was linked firmly with one man – Sir William Lyons. His was the personality, the drive and the vision which turned Jaguar from a small company into a world-renowned marque. Quite simply, without Sir William there would have been no Jaguar XJ6, and no Jaguar story to tell. Without Sir William, indeed, this type of car might never even have been invented.

## LANCASHIRE BEGINNINGS

William Lyons was born in Blackpool in 1901, and raised in that seaside town, although his first job was as an engineering apprentice at Crossley Motors, Gorton, Manchester. This seems to have bored him, for very soon he abandoned that enterprise, returned to Blackpool to hang around his family's piano renovation workshops, and began to dream of running his own business.

Soon he started riding motorcycles, then joined a local garage business selling cars – Morris, Rover or Sunbeam models. Shortly, though, the Lyons family moved house to King Edward Avenue in Blackpool, where William realized that across the road an older man called William Walmsley was developing a tiny business to manufacture lightweight motorcycle sidecars, called Swallows.

William Lyons bought a Swallow sidecar in 1921, saw the opportunity to get involved with a business which he might enjoy, badgered his parents to put up some money to buy him a partnership in it, and moved into a small factory in Bloomfield Road, Blackpool in 1922. He was only twenty years old.

Right from the start, Lyons proved to have a good eye for a line, and soon took over the shaping of Swallow's products. In the next decade, Swallow not only produced more and more sidecars, but branched out as a builder of special bodies for cars. The first such rebodied car was the Austin Seven Swallow of 1927, which was rapidly followed by the Morris Cowley Swallow. In the next few years, Swallow bodies would also find their way on to Swift, Standard and Wolseley chassis.

## THE MOVE TO COVENTRY

By 1928, the Blackpool-based business was booming. The factory itself was already overcrowded, and in addition William Lyons (if not William Walmsley) had his sights set on greater things. Blackpool, and sidecar bodies, was now too provincial a scenario for him – he wanted to be at the centre of the motor industry, to expand, and to build a larger business.

The deciding factor was that Swallow received an order for 500 Austin Seven Swallows from Henlys Ltd of London. To cope with this, and with Swallow's existing business, the company had to move, and a new home was found in Whitmore Park, Foleshill, a northern suburb of Coventry. The factory itself was a disused shell-filling factory which had been erected during World War I, and the move took place at the end of 1928.

Before long, Swallow Coachbuilding was making up to fifty car bodies every week, in spite of the troubled economic climate which afflicted the UK at this time. The unmade access road outside the factory was eventually surfaced, and named Swallow Road. The company made its first appearance at the annual Olympia Motor Show in October 1929, which effectively meant that the company had become an accepted part of the 'establishment' motoring scene.

## SS – A NEW MARQUE IS BORN

The next step was for Swallow to start building 'own-brand' cars, and in 1931 an exclusive deal with John Black of Standard made this possible. Standard agreed to supply special Standard Sixteen six-cylinder chassis for Swallow to clothe with a rakish two-door coupé body style, the result being the birth of the SS car.

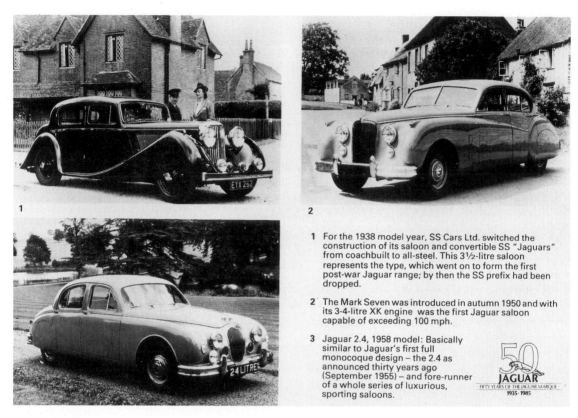

1 For the 1938 model year, SS Cars Ltd. switched the construction of its saloon and convertible SS "Jaguars" from coachbuilt to all-steel. This 3½-litre saloon represents the type, which went on to form the first post-war Jaguar range; by then the SS prefix had been dropped.

2 The Mark Seven was introduced in autumn 1950 and with its 3-4-litre XK engine was the first Jaguar saloon capable of exceeding 100 mph.

3 Jaguar 2.4, 1958 model: Basically similar to Jaguar's first full monocoque design – the 2.4 as announced thirty years ago (September 1955) – and fore-runner of a whole series of luxurious, sporting saloons.

*When the Jaguar name reached its fiftieth anniversary, the company issued this composite picture, showing three generations of Jaguar car, all of which had been styled by Sir William Lyons. All were 'firsts'. Top left the original late-1930s SS-Jaguar, top right the Mark VII (the first saloon to use the XK engine), and bottom left the 2.4 litre (the first to have a monocoque body/chassis design).*

The first point to be settled, of course, is the meaning of 'SS'. Was it Standard Swallow, Standard Special, Swallow Sidecars, or Swallow Special? I do not know – and as far as I know Sir William Lyons never spelt it out, though 'Standard Swallow' was always favourite. What *is* important is that the initials were incorporated in a new company (SS Cars Ltd) and that they came to be linked with the cars from Coventry which had been personally styled by the company's thrusting partner.

There were two SS cars on display at Olympia in October 1931, the six-cylinder SS1, and the much smaller SS2, which was based on the rolling chassis of the Standard Little Nine. Neither was yet quite ready to go on sale, though the orders soon began to roll in. At launch, the rakish SS1 immediately set a precedent which later Jaguars followed – it looked much more expensive than it was. Press comments were that it looked smart enough to cost £1,000 (a *very* high price for 1931), though it actually went on sale at £310.

It would be fair to say that William Lyons' body styles flattered the chassis that they clothed, for at this stage there is no doubt that cars looked a lot faster than they actually were. Standard's chassis were sound enough, but the side valve engines had little to recommend them except reliability. Even so, SS car sales built up rapidly, and by the mid-1930s not only the coachbuilding activity for other manufacturers, but the manufacture of sidecars, had to cease. For William Lyons, however, this was only the beginning.

The word 'Jaguar' first came on to the scene in 1935 and, looking back, this was certainly a year of destiny for the business. Not only was the SS-Jaguar motor car born, but William Walmsley pulled out of the business, the first overhead-valve engines were used, and SS Cars Ltd 'went public'.

SS Cars Ltd was floated in January 1935, with a capital of £250,000, at which point William Walmsley walked away from the business which he had founded in the early 1920s. William Lyons, now in sole charge, immediately pressed ahead with his next enterprise, which was to produce a series of new cars with his own design of chassis, and with specially developed engines. This was the point at which William Heynes joined SS Cars as Chief Engineer, and at which Standard started building overhead-valve engines which had been developed by Harry Weslake's business. The new cars, launched at Olympia in the autumn of 1935, needed a new name, and although there is a strong legend that William Lyons chose Jaguar from a long list of animal names which he consulted, it actually seems to have been one of several names suggested to him by his advertising agents.

The choice of 'Jaguar', however, could not have been more satisfactory, for it suggested grace, speed and agility, and was right for the sort of car William Lyons produced. At first, for 1936, there were 1.5-litre and 2.5-litre SS-Jaguar models, but soon they were also joined by the 3.5-litre range, and by the wickedly attractive SS100 two-seater sports cars.

---

### William Heynes (1903–1989)

Before Bill Heynes started work for William Lyons in 1935, the company had no real engineering department. Lyons hired him to change all that. By the 1950s Jaguar's technical resources had mushroomed, the company's twin-cam XK engine was famous, and its racing reputatin was established.

William Munger Heynes was born in Leamington Spa, near Coventry, attended Warwick school, and if his masters had got their way, he might have become a surgeon. Instead he joined Humber Ltd as an apprentice, ran the technical department from 1930, and remained with that firm until April 1935, when he joined SS Cars Ltd.

His speciality at this time, was chassis and suspension design, which dovetailed neatly with William Lyons' love of body styling. As Chief Engineer at SS, and with only one draughtsman to assist him, his first job was to finalize the new SS-Jaguar models, after which he masterminded the design of the overhead-valve engines which served SS-Jaguar, later Jaguar, until 1950.

Before retiring from Jaguar in 1969, he became the Company's Engineering Director and Vice-Chairman. Although Sir William Lyons would never have admitted this, he looked on Heynes as his most important right-hand-man for many years, for it was under Heynes that classics like the XK engine, the XK120 sports car, the Mk VII, the 2.4-litre saloon range, the famous C-type and D-type racing cars, the E-type, and the XJ6 were all developed. The fabulous V12 engine, too, was well on its way to launch before he retired at the age of 66.

Heynes was always a dapper little man, in latter years addicted to what his drawing-board staff called 'the 4B pencil redesign job', who ran his engineering departments like a personal fiefdom. Subordinates like Claude Baily, Walter Hassan and Bob Knight were all experts in their own way, but Heynes, who could be stubborn as well as dismissive, always got his way in the end.

When war broke out in 1939, not only was SS-Jaguar well on the way to strangling the sales of rival companies like Alvis, Armstrong-Siddeley, Riley and Triumph, but it was still expanding steadily. The Foleshill factory had been expanded, but was still crammed full, the saloon bodies used were now made entirely from pressed-steel panels, and the powerful 3.5-litre engines had endowed the SS100 sports car with a top speed of 100mph (160kph). In 1937 production topped 3,000 for the first time, and in 1939 (in spite of the onset of war) it reached 5,378.

During the war, SS built a variety of military equipment, but when his management team was obliged to carry out overnight fire-watching duties at the factory, William Lyons made sure that they were also kept busy planning for the company's future. *This* time the legend is also the true story – the famous twin-overhead-camshaft series of XK engines was conceived in those dark days and nights when Britain, and its citizens, were quite literally fighting for their lives.

## JAGUAR TAKES OVER FROM SS

Before and during World War II, the initials 'SS' had taken on sinister connotations, when applied to Hitler's feared and hated storm troopers, so it was no surprise when, in 1945, William Lyons decided to rename his company. Henceforth the cars would become

*For many years before Sir William Lyons took over Daimler, to add to the Jaguar business, it was making successful cars. The Autocar's Sports Editor, S.C.H. Davis, drove this smart 2.5-litre example in the 1939 Monte Carlo Rally.*

Jaguars, and the company which would build them was to be Jaguar Cars Ltd.

In the next ten years the new company, in which William Lyons personally had majority control, seemed to be one constant blur of activity and innovation. Between 1946 and 1956, production moved up from 1,132 to 12,152 cars a year – a 1,000 per cent leap – while annual profits rocketed from £22,852 to £326,676; most of those profits were reinvested, for Jaguar seemed to be expanding continuously.

To give a flavour of the way the company changed in those years, I need only list the highlights:

| 1946 | Post-war car production began |
| 1948 | XK twin-cam engine introduced |
| | XK120 sports car introduced |
| | Mk V saloon introduced |
| 1950 | Mk VII (twin-cam engined) saloon introduced |
| 1951 | Jaguar's first victory in Le Mans 24-Hour race |
| 1951/52 | Factory moved from Foleshill to Browns Lane, Allesley |
| 1954 | Introduction of sleek D-type racing sports car |
| 1955 | Launch of first-ever monocoque saloon model, the 2.4 |

Was it any wonder that William Lyons, the human dynamo behind all these enterprises, became *Sir* William in the New Year's Honours List of January 1956?

Sir William had always been an ambitious man, but his constant urge to expand the company was also tinged with fiscal caution. He rarely splashed out on publicity stunts, his careful monitoring of expenditure, and his abhorrence of waste, was well known, and he was fair rather than generous with salaries. At a time when other businessmen were getting rich, quick, he certainly made money, but also gave great value.

---

### Browns Lane

All SS cars, all SS-Jaguars, and early post-war Jaguars were produced in the Foleshill factory in the northern suburbs of Coventry. By 1950, that factory was bursting at the seams, and William Lyons began to look round for more factory space. Coventry, at the time, was going through something of an industrial upheaval, not only because wholesale rebuilding to repair World War II damage was well under way, but also because several of the original aero-engine shadow factories were becoming available.

The government's shadow factory scheme was launched in 1936, as a way to boost the rearmament programme, and particularly to get the motor industry involved in setting up and managing new aero-engine factories. No sooner was the first wave of factories up and running, than the government initiated phase two, the result being that Daimler built 'Shadow No. 2' in Browns Lane, Allesley (to the west of Coventry). This building began by producing Bristol Hercules radial engines during the war, but by the early 1950s it was closed down, Daimler had no further use for it.

After a great deal of negotiation with government officials, Mr Lyons was able to buy (not lease) the factory, began moving his assembly facilities into place during 1952, and completed the move by the end of that year. Once the move was completed, the Foleshill factory was sold off, and all links with the SS and SS-Jaguar days were dissolved.

Browns Lane has been the home of Jaguar ever since then, though it has been modified, expanded, rejigged, and persistently modernized ever since. Even as recently as 1991, the city council opened a new road linking the Allesley bypass with Coundon, in which a major feature was a new roundabout, and direct access to a new 'back door' to Browns Lane.

## The Birth of the XK Engine

Jaguar's famous XK engine, which powered every six-cylinder XJ6 built from 1968 to 1986, was conceived during wartime Sunday night fire-watching sessions at Foleshill from 1943 to 1945, was unveiled in 1948 as the XK120 sports car's power unit, and was the *only* Jaguar passenger-car engine produced from 1950 to 1971. Although the second-generation XJ6 of 1986 used the new AJ6 engine, the old-type XJ6 was still being made in the early 1990s to power the long-running Daimler limousine.

William Lyons wanted his post-war engines to be powerful, to *look* powerful, but to be smooth, refined, and reliable. This was a difficult, but not impossible, brief for William Heynes, Claude Baily and Walter Hassan to match, but history proves that they achieved the standard.

The first projects encompassed a family of four-cylinder and six-cylinder engines. The very first prototype was the twin-cam four-cylinder 1,360cc XF unit, while the second was the BMW-style cross-pushrod XG, which was a conversion of a Standard Flying 14/Jaguar 1.5-litre engine.

The *true* forerunner was the XJ, in which the four-cylinder version had a bore of 80.5mm and a capacity of 1,996cc, the six an 83mm bore and 3,182cc, while there was a common cylinder stroke of 98mm. This design matured as the XK when the stroke of the six-cylinder engine was lengthened to 106mm to provide more torque, and the definitive 3,442cc engine was born.

Production, on a limited scale, began in 1949, at which point the XK engine was the world's *only* twin-overhead-camshaft engine in series production. By the time the XK engine found its way into the first XJ6 saloon, it had already powered cars as diverse as the XK sports cars, the Mk VII saloons, the D-type racing sports cars, and the Alvis Scorpion armoured military vehicle. It was, and still is, a legend in its own lifetime, and has been fitted to hundreds of thousands of Jaguars.

When the Foleshill factory finally ran out of space in 1950 (and when government planners would not allow him to erect new buildings on wasteland alongside the existing plant), he cast around for a new home and found it just a couple of miles away, in Allesley. Daimler's No. 2 Shadow Factory (erected at the end of the 1930s for the manufacture of aero engines) was redundant, so a deal was quickly done, and Jaguar's facilities were gradually, and progressively, moved during 1951 and 1952.

Three major developments put Jaguar into the headlines in this period – the XK engine, the XK sports car, and success in motor racing. The XK engine – the fire-watching design – was actually developed by William Heynes, Claude Baily and Walter Hassan as a family of four-cylinder and six-cylinder units, though only the six-cylinder engine ever went into production, this being the heart of every important Jaguar car built in the 1950s and 1960s. It started life as a 3.4-litre unit, but was eventually built in 2.4, 2.8, 3.0, 3.4, 3.8 and 4.2-litre form. The XK was Britain's first ever series-production twin-cam engine, which looked as good as it performed. It would do a great job for more than 40 years.

The XK sports car family – XK120, XK140 and XK150 – started life as a 'short-term' XK120 project in 1948, to give exposure to the new engine, but became so successful that new production tooling was laid down to give the cars a longer life. That life eventually ran from 1948 to 1957, and Jaguar's reputation rode on the back of it.

Thirdly, of course, there was the motor racing programme, which to William Lyons meant an assault on the Le Mans 24-Hour race. A tentative entry with XK120s in 1950 led to new XK120 C (C-type) cars being designed for 1951. Those cars won in 1951 and again in 1953. Their successors, the charismatic D-types, arrived in 1954, and went on to win at Le Mans in 1955, 1956 and 1957.

## TECHNICAL ADVANCE
## IN THE 1950s

Until the mid-1950s, Jaguar's production cars were all built on the basis of sturdily engineered separate chassis frames with independent front suspension and beam rear axles, topped by separate steel bodyshells. In the case of the XK sports cars, Jaguar actually assembled its own shells from panels supplied by outside specialists.

In the meantime, the British motor industry was gradually, but inexorably, changing over to the use of monocoque (unit-construction) body/chassis units, where there was no separate frame. At this time very few companies were capable of building such shells, which required complex and expensive jigs and assembly fixtures.

William Lyons and his technical chief William Heynes could see the obvious attractions – structures were usually lighter, stiffer, and more compact – but at the same time they could see the obvious snag, which was that tooling costs were much higher. Making money from a new design built around a unit-construction shell would mean selling many more cars, and having those cars in production for a lengthy period. Since William Lyons had already decided that this was the correct policy for his company, he decided to go ahead.

The first unit-construction Jaguar to go on sale was the 2.4-litre saloon, and except for the E-type sports car of 1961–75 (which was a special case), all future Jaguars would follow the same route. But Jaguar could neither design nor manufacture its own new-generation monocoque shells. Its only course was to rely on one of the independent concerns which could cope – the Pressed Steel Co. Ltd of Cowley.

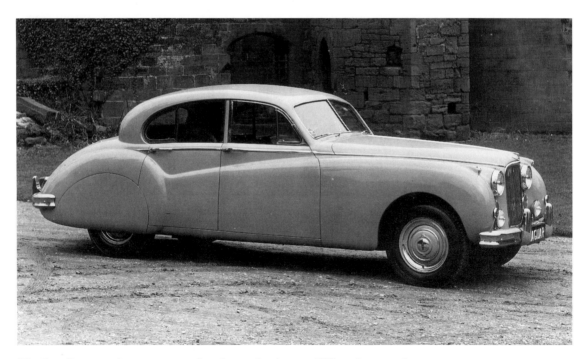

*The first Jaguar saloon car to use the advanced twin-cam XK engine was the Mk VII model. This is the slightly face-lifted and more powerful Mk VIIM version. The XK engine was still used, and still being developed, when the first XJ6 was produced in 1968.*

In the next ten years, Jaguar's business expanded, and the model range became increasingly complicated. Not only that, but the only link between 1955 and 1965 was the six-cylinder XK engine itself. The following list of products tells its own story:

**1955**

XK 140 sports car – separate chassis, 3.4-litre engine
Mk VIIM saloon – separate chassis, 3.4-litre engine
9,900 cars were built in that year

**1965**

E-type sports car – monocoque/tubular chassis, 4.2-litre engine
Mk II saloons – monocoque, 2.4/3.4/3.8-litre engines
S-type saloons – monocoque, 3.4/3.8-litre engines
Mk X saloons – monocoque, 4.2-litre engines
Daimler V8–250 saloon – monocoque, 2.5-litre V8 Daimler engine
24,601 cars were built in that year.

By the 1960s, in fact, Jaguar had diversified considerably. After taking over Daimler in 1960 it not only carried on building Daimlers at the Radford factory, but it developed a Daimler-engined version of a Jaguar saloon. The last of the separate-chassis saloons – the Mk IX (the last derivative of the Mk VII) – was produced in 1961, and had been replaced by the gargantuan monocoque Mk X saloon.

There was more complication to come, for in 1966 yet another model line, the 420, would appear, and by that time it was quite easy (and easily justified) to accuse Jaguar of having too many models. The XJ6 model was specifically developed to rectify this problem.

## 2.4 AND SUCCESSORS

If you agree that Jaguar was trying to build too many different cars in the 1960s, it is very easy to find the culprit, which was the original 2.4-litre saloon of 1955. Not only

---

### Daimler's Radford Factory

Daimler was Britian's oldest car maker, beginning its long and illustrious industrial career in a converted mill in Sandy Lane, Coventry. Even in the early days this was a very inconvenient factory for the assembly of cars, so in 1908 the company bought an adjacent site in Radford, to the north of the city centre, but did not build on it before it was taken over by BSA.

BSA built a munitions factory at Radford during World War I, and in the 1920s and 1930s this became Daimler's principal manufacturing plant. In the 1930s it not only built Daimler cars but BSA and Lanchester cars, along with Daimler buses, while World War II also saw it producing Daimler military scout cars and armoured cars.

Daimler's first shadow factory was built alongside the original Radford plant in 1937 – though in the 1950s this sizeable facility was sold off to Standard-Triumph to establish a transmissions factory; it is now back in Jaguar's hands. When Jaguar bought Daimler in 1960, it immediately became Radford's owner which, at a stroke, meant that Jaguar had twice as much floor space as before.

During the 1960s, Daimler's rambling business at Radford, which had not only embraced a diverse range of car assembly, buses and military vehicles, but engine and transmission manufacture for the whole range, was completely re-equipped.

By the late 1960s, all Jaguar-Daimler car assembly was centred on Jaguar's Browns Lane factory, while Radford was given over to engine manufacture (XK engines, later V12 engines, and – from the 1980s – AJ6 engines), and to gearbox manufacture.

*Just before Jaguar took over Daimler, Daimler produced this glass-fibre bodied two-seater sports car, the SP250. The styling was strange, but it hid a fine V8 engine.*

the Mk II models of 1959, but the S-types and 420s which followed were all developed from this one design. The 2.4-litre was launched in 1955, and was soon followed by the larger-engined 3.4 of 1957. The Mk II models of 1959 featured restyled cabins, and a choice of three engines – 2.4, 3.4 and 3.8-litres.

Next was the Daimler-badged V8–250 saloon, which combined the Daimler SP250 sports car's V8 engine with the Jaguar Mk II monocoque. To follow up, the S-types, launched in 1963, used modified versions of the monocoque, this time with lengthened tails and independent rear suspension.

Finally, in 1966, Jaguar launched the 420 saloon, which took the S-type theme a stage further, this time with a new and squared-up nose, and a 4.2-litre engine. To add a bit of further glitz, there was also to be a Daimler version of this car, badged as a 'Sovereign'. This frenetic activity would have been more logical if one new model had replaced an old one, but at Jaguar in the 1960s this never happened. The S-type family was added to the Mk IIs, and the 420 was added to both of them.

What happened next – what had to happen, indeed – is ideally summed up in

*After the Mk II model, Sir William Lyons moved perceptibly towards the late-1960s XJ6 style when he shaped the S-type of 1963. There was to be one further evolution on this theme, when the 420, which was like this car but with a squared-up nose appeared in 1966.*

*The sensational E-type of 1961 had evolved into this broader, heavier and V12-engined Series III type by the early 1970s. When it was finally dropped, Jaguar decided not to produce a direct replacement, but chose to design the much larger four-seater XJ-S instead.*

*The very old, and the new – a mid-1980s shot showing how motor cars had developed in ninety years, for this is a direct comparison between an 1898 Daimler and a 1986 Daimler Double-Six.*

Jaguar historian Andrew Whyte's own words:

One thing was certain: Jaguar was beginning to make too many models. The E-type range was doing very well, thank you, exceeding 200 units a week at the peak of its appeal. The saloon range, however, needed slimming down. Interim models ... would soon need to be supplemented and/or succeeded by Lyons' ultimate Jaguar motor car. This was intended to be the car with the world's quietest and most refined riding qualities, bar none, and retain Jaguar's famous attributes of performance, controllability, comfort and value for money.

By the mid-1960s the XJ6 project was under way.

# 2 XJ6 – The First Five Years

Although first thoughts on a new and rationalized range of Jaguars took shape early in the 1960s, the first serious design work started in 1964. It would take four years before the definitive car, to be badged XJ6, was ready to be unveiled.

In the 1960s, as in the previous thirty years, Jaguar went about the development process in its own distinctive way. To Sir William Lyons, the looks of the car were at least as important as the engineering that turned a shape into a working car, so the styling of new models always took priority.

First of all, Sir William would get together with his engineering Vice-Chairman, Bill Heynes, to agree on the bare bones, the platform, of the new car's layout – its wheelbase, its tracks, and (very important, this) its tyre size. Sir William could then get on with the styling of the shell, and

*Two generations of XJ saloon prove how little the general proportions of these elegant cars have changed in more than twenty years. The red car is one of the very first XJ6s ever built, a 1968 example, while the blue car is a 1991 model Jaguar Sovereign.*

the engineering project work marked time for a while.

Compared with almost every other company in the business, Jaguar's styling evolved in a unique manner. Sir William was not only the Chairman and controlling director of the business, but he was also the company's stylist. But not with sketches, and not with perspective drawings – for Sir William a new model had to take shape in the solid, in full size. This method, incidentally, could never have succeeded without a dedicated department which understood his needs, and his methods.

This process, established with the launch of the original SS1 in 1931, revolved round a forceful character called Fred Gardner. Fred, who had joined Swallow at Foleshill in 1929, was the son of a Liverpool seaman and originally came to run SS's 'wood shops', which not only produced body frames, but the fine veneers used to cover dashboards, and for door cappings.

Gardner ruled his departments with a rod of iron for more than forty years; he could charitably, and accurately, be described as a 'real character'. He is also reputed to be the *only* Jaguar employee or director whom Sir William habitually addressed by his Christian name . . .

Tucked away in a corner of the works, Gardner had a small experimental department which was devoted to building full-scale mock-ups for Sir William. Jaguar's Chairman spelt out his requirements, drew three-dimensional pictures in the air with his hands, then left Gardner's team to interpret his needs. They would rapidly build up a wooden skeleton, clothe it in hastily hand-beaten metal panels, and offer it to Sir William for inspection.

The first effort was rarely even close to the final product. Sir William, who found time to visit this tiny department every day, demanded constant changes while the mock-up was being built and then, from time to time, asked for the shape to be placed outside for a more general inspection. You might think, and you would be correct, that this was a hit-and-miss process. You might also suspect that first attempts were rarely acceptable, and in the case of the XJ6, Jaguar's own archives prove the point. Sir William's reputation as an instinctive styling genius was often overstated by his many disciples. He was *not* infallible; when he was starting work on a new shell his early whims often produced ungainly shapes, but at least he was realistic, had a good eye for what looked right and what looked wrong, and would invariably arrive at the correct solution, even if it took a long time to get there.

[When I was working at Jaguar in the late 1950s, as a young man, a successor to the Mk VIII was being developed in Fred Gardner's secret department. Known officially as the 'new Mk Nine', it took so long to be finalized that design staff jocularly referred to it as the 'Mark Time' – but not if Sir William was within earshot! It took *so* long, in fact, that it had to be called Mk X, for a Mk IX had already intervened by that time . . .]

## XJ6 – WHAT SIZE AND TYPE OF CAR?

In the early 1960s that well-known phrase 'product planning' was not so well known at Jaguar. When a new model was being defined Sir William would talk to his sales staff and his engineers, make up his own mind about the bulk and the shape of the car, but would stand back and listen carefully when engine sizes and options were proposed.

The car we now know as the XJ6 was intended to take over from all existing Jaguar and Daimler saloons, including the large Mk X model. This was going to be a demanding brief, for in the early 1960s the range stretched from the 112bhp 2.4-litre

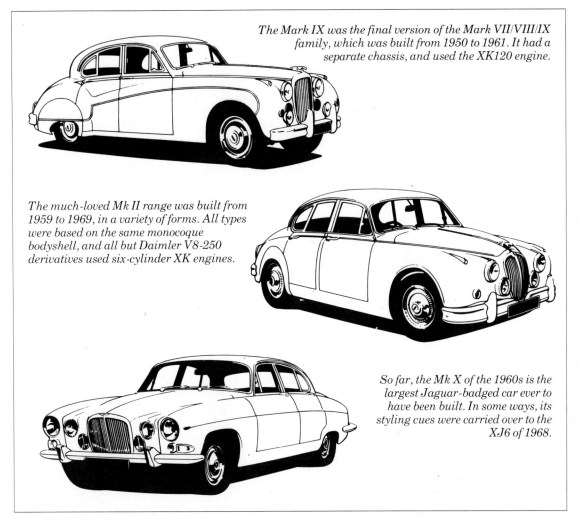

*The Mark IX was the final version of the Mark VII/VIII/IX family, which was built from 1950 to 1961. It had a separate chassis, and used the XK120 engine.*

*The much-loved Mk II range was built from 1959 to 1969, in a variety of forms. All types were based on the same monocoque bodyshell, and all but Daimler V8-250 derivatives used six-cylinder XK engines.*

*So far, the Mk X of the 1960s is the largest Jaguar-badged car ever to have been built. In some ways, its styling cues were carried over to the XJ6 of 1968.*

Mk II to the 265bhp 4.2-litre Mk X. By the time the XJ6 was ready for sale, it had to provide a successor for all these cars:

1   Mk II saloons, built on an 8ft 11.4in (2,728mm) wheelbase, were 15ft 1in (4,597mm) long and had a choice of 2.4, 3.4 and 3.8-litre XK engines.
2   The Daimler 2.5-litre (V8–250) saloon, like the Mk II, but with a 2.5-litre Daimler V8 engine and 140bhp.
3   S-type saloons, which had the Mk II wheelbase, but which had independent rear suspension, were 15ft 8in (4,775mm) long,

and had a choice of 3.4-litre or 3.8-litre XK engines.
4   420 saloons, which were Mk II derived, but had only 4.2-litre XK engines. There was a Daimler Sovereign derivative of this model.
5   420G saloons (renamed from Mk X), which had very large shells on a 10ft 0in (3,048mm) wheelbase, were 16ft 10in (5,131mm) long, and had 4.2-litre XK engines.

To do this, Sir William wanted to style a monocoque body shell with a more roomy interior than the old Mk II/S/420 cabin, but a rather smaller shell than that of the Mk X. He chose a 9ft 0.75in (2,762mm) wheelbase (which was only slightly longer than that of the old Mk II model), but settled on wheel tracks of 4ft 10in (1,473mm), which were the same as those of the MK X/420G, and three inches (76mm) wider than those of the Mk II/S/420 model. This allowed the cabin width to be considerably increased, as one comparative figure shows. On the original Mk II the 'across-shoulders' dimension in the front seat was 52in (1,321mm), while in the new XJ6 it was 54in (1,372mm). The Mk X measured 57.5in (1,461mm), but was a massive 6.5in (165mm) wider externally than the XJ6. The new car, in other words, had an improved and more spacious interior package.

*The original XJ6 style needed very little modification over the next eighteen years, for it was an elegant, spacious, and capable saloon. The styling, by Sir William Lyons himself, featured four headlamps, a slight suggestion of flared wheelarch over the rear wheels, and a very wide and purposeful stance.*

*Sir William liked to see his cars shaped 'in the round', in wood, and spent hours pacing round them, to ask for a line to be accentuated here, or a trifle to be shaved off there. The XJ6 was elegant from every angle, and has always been a tribute to the sure 'eye' of its designers.*

The engine bay was more commodious than on the Mk II models, for room had to be found for the existing XK engine, and for the bulkier V12 which was also on the way. In 1964, however, when styling work began, the layout of the V12 had not been finalized.

Getting the shape right took a little time. First attempts produced a very curvaceous car with a distinctive kick-up of the wing crown (or waist) line above the rear wheels, with an angled four-headlamp installation, and a rounded tail owing something to the shape of the E-type. There was a bulge in the

bonnet, and a rather anonymous grille/radiator opening. This was soon refined with a smoother through line, a cut-off tail, a smooth bonnet, and a squared-up front grille flanked by four separately mounted headlamps. This front-end style was a reversion to that of the 420 saloon (which was to be launched in 1966), though a considerable number of detail changes were made before the entire shape was frozen, ready to be committed to tooling by Pressed Steel. The finalized shape, of course, was yet another classic by Sir William, one which had been

## XJ6 Series I (1968–1973)

**4.2-litre** (UK-market specification).

### Layout
Unit-construction monocoque five-seater, front engine/rear-drive, sold as four-door saloon

### Engine

| | |
|---|---|
| Block material | Cast iron |
| Head material | Cast aluminium |
| Cylinders | 6 in-line |
| Cooling | Water |
| Bore and stroke | 92.07 × 106mm |
| Capacity | 4,235cc |
| Main bearings | 7 |
| Valves | 2 per cylinder; DOHC operation |
| Compression ratio | 9.0:1 (8.0:1 optional) |
| Carburettors | 2 SU HD8 |
| Max. power (DIN) | 173bhp @ 4,750rpm |
| Max. torque | 227lb/ft @ 3,000rpm |

### Transmission

| | |
|---|---|
| Clutch | Single dry plate; diaphragm spring; hydraulically operated |

### Internal gearbox ratios

| | |
|---|---|
| Top | 1.00:1 |
| 3rd | 1.389:1 |
| 2nd | 1.905:1 |
| 1st | 2.93:1 |
| Reverse | 3.378:1 |
| Final drive | 3.54:1, later 3.31:1 |

Optional overdrive, (0.779:1 ratio) and 3.77:1 final drive ratio

Optional Borg Warner automatic transmission, with torque converter

### Internal ratios

| | |
|---|---|
| Top | 1.00:1 |
| 2nd | 1.45:1 |
| 1st | 2.40:1 |
| Reverse | 2.09:1 |
| Maximum converter multiplication | 2.0:1 |
| Final drive | 3.54:1, later 3.31:1 |

### Suspension and steering

| | |
|---|---|
| Front | Independent by coil springs, wishbones, anti-roll bar and telescopic dampers |
| Rear | Independent, by double coil springs, lower wishbones, fixed length drive shafts, radius arms and twin telescopic dampers |
| Steering | Rack-and-pinion, power-assisted |
| Tyres | E70 205VR-15in |
| Wheels | Pressed steel disc |
| Rim width | 6.0in |

**Brakes**

| | |
|---|---|
| Type | Disc brakes at front and rear |
| Size | 11.8in diameter front discs, 10.4in rear discs, with vacuum servo assistance |

**Dimensions (in/mm)**

| | |
|---|---|
| Track | |
|   Front | 58/1,473 |
|   Rear | 58.3/1,481 |
| Wheelbase | 108.8/2,763 |
| Overall length | 189.5/4,813 |
| Overall width | 69.7/1,770 |
| Overall height | 53/1,346 |
| Unladen weight | 3,703lb/1,676kg |

**Long-wheelbase 4.2-litre, specification as for normal-wheelbase car except for:**

**Dimensions (in/mm)**

| | |
|---|---|
| Wheelbase | 112.8/2,865 |
| Overall length | 194.7/4,945 |
| Unladen weight | 3,793lb/1,720kg |

**2.8-litre specification as for shorter-wheelbase 4.2-litre except for:**

**Engine**

| | |
|---|---|
| Bore and stroke | 83 × 86mm |
| Capacity | 2,792cc |
| Compression ratio | 9.0:1 (optional 8.0:1 or 7.0:1) |
| Max. power (DIN) | 140bhp @ 5,150rpm |
| Max. torque | 150lb/ft @ 4,250rpm |

**Transmission**

| | |
|---|---|
| Final drive ratios | 4.27:1 (manual), 4.55:1 with overdrive, 4.27:1 with automatic transmission |

**Dimensions**

| | |
|---|---|
| Unladen weight | 3,388lb/1,534kg |

achieved through his artistic eyes, and through the patience of Fred Gardner's craftsmen.

# ENGINES AND TRANSMISSIONS

In 1965, Jaguar was building too many different XK engines. The design was already seventeen years old, and now that the 4.2-litre version had been revealed, there was no more 'stretch' and little more development locked inside.

For the new car, therefore, first proposals were to use only one version of the XK engine – a 3-litre with a shorter stroke and a shallower block than the 4.2-litre – along with the newly designed 5.3-litre V12 unit, and several derivatives of that new engine. The description of the V12, and of the 'slant-6' and the 'V8' derivatives of it, belongs to the next chapter.

There was no chance of the new V12, or its

*In eighteen years the general proportions of the XJ6 would never need to be changed, though this original 1968 car would eventually give way to a longer wheelbase type, and there would be a retouched cabin shape in the 1980s. There was a distinct family resemblance to previous Jaguars, especially in the layout of the cabin's rear quarters and window profiles.*

related V8, being ready by the time the XJ6 went into production, so the initial cars had to be produced with XK power only. The original strategy did not even pass the early prototype stage. Although a 3-litre engine was indeed built and tested, it was soon rejected for the same reason that the *original* XK engine of 1946–7 had been turned down – there was plenty of top-end power but a lack of low-speed torque.

At the same time, the sales force pointed out that there was an important 'tax break' for engine sizes, at 2.8 litres, in certain European countries, so a change of course was called for. Instead of a 3-litre engine, the first XJ6s would have a choice of 2.8-litre or 4.2-litre units. The only other imposed change because of this decision was that a shapely bonnet bulge was needed, to give clearance over the taller 4.2-litre engine.

As Jaguar had recently revised its choice of transmissions, this caused the designers very little worry. Most cars were expected to have automatic transmission, these being the well-proven Borg Warner three-speed unit (Type 35 for the 2.8-litre, Model 8 for the 4.2-litre). Manual transmission cars had the latest all-synchromesh four-speed unit (which had been introduced as recently as 1964, at the time of the launch of the 4.2-litre engine) with a Laycock overdrive once again tucked neatly in behind it. A limited-slip differential was standard in the rear axle of the 4.2-litre cars.

## STRUCTURE, CHASSIS – AND CHARACTER

As every enthusiast now knows, the three outstanding features of the XJ6 were its stiff structure, its silky refinement, and its unique combination of ride and handling. In 1968, quite simply, no other car in the world could touch it. Ever since Jaguar had introduced its first unit-construction car (the 2.4

*No detail was missed in styling a new Jaguar. From the rear, the original-type XJ6 shows off the careful integration of sheet metal shapes, bumpers, badges and tail lamps. On this car only the mud flaps were not originally standard.*

saloon of 1955), it had been known for designing stiff shells and agile cars, but the XJ6 surpassed its ancestors, and set completely new standards. Rolls-Royce and Mercedes-Benz, for sure, were immediately outclassed.

All these features were linked to the design of the monocoque. As usual, Jaguar styled the car and schemed its structure before handing over to Pressed Steel of Cowley for the detail engineering to be completed on its behalf. The result was a shell which, when tested, was found to be lighter than expected and to have a torsional stiffness of 8,500lb ft/degree, a very high figure which was the best so far achieved by any Jaguar.

More important than this stiffness, however, was the fact that the production car was quite astonishingly refined. There was an almost complete absence of what engineers used to call 'road-excited body noise', the XK engine was almost silent in operation unless thrashed to peak revs all the time, while 'bump-thump' from the suspension had virtually been banished. Some of this achievement was to the credit of Bob Knight's team; Bob was then Jaguar's Chief Development Engineer, but after Bill Heynes retired became the company's Chief Chassis Engineer. Some of it was due to the way Tom Jones and his chassis designers applied all their previous experience to the new project, forgetting nothing they had learned on earlier models, and adding new improvements along the way.

Although both independent front and rear suspensions were logical developments of those already being used on existing Jaguars, they were more effective than before. A side-to-side look at layouts showed what had been done – at the front end, for instance, there was a box-section cross member in place of the Mk X's forged beam, while the suspension dampers on the XJ6 had been mounted outboard of the coil springs, not only to make them more effective, but to allow longer units to be installed.

## The XJ6 Independent Rear Suspension

The first Jaguar road car to have independent rear suspension was the famous E-type, but forerunners of this layout had been built in the 1950s, and several refinements had already been made before it was adopted for the XJ6 of 1968.

When Jaguar started racing the D-type in 1954, this car had beam axle rear suspension, but after the drivers complained of lack of traction, two further systems were designed and tested – one being a de Dion suspension, the other an independent layout – though neither system was adopted.

The independent layout, which featured twin coil spring-over-telescopic dampers at each side, a fixed length drive shaft, and complex lower wishbones at each side, was then further refined for the E-type project, was always a part of that design, and was adopted for production 1961.

Its first public appearance was in E2A, the 1960 Le Mans car, when there were no fore-and-aft radius arms, but the production type, complete with a massive pressed steel subframe and rubber mountings to the shell, was revealed in March 1961. To follow the E-type, the Mk X (later renamed 420G), the S-type, and the 420 models all had the same type of suspension.

Some of the improvement, undoubtedly, was due to this sort of attention to detail, which went into the layout of suspension mountings, and into chassis tuning. The advance started with the tyres, which were Dunlop E70VR15in radials riding on six-inch wheel rims. Inboard of these were front and rear suspension subframes which either damped out road vibrations, hid it from the passengers, or supported rubber/plastic mountings which did their own magic. Most of all, however, it was Bob Knight's team, which included Norman Dewis as a redoubtable test engineer, who went to a great deal of trouble to work away at the suspension car took place while Jaguar was finding itself a business partner. Rumours had begun to spread in the 1960s that Jaguar was looking for a partner, not because it needed financial support, but because Sir William saw this as a way of underpinning his company's future.

We will never know the actual reasoning behind Sir William's move, and the identity of other suitors (except that he had appa-rently decided that his company could not survive for long without him, if and when he decided to retire), but the fact is that in the summer of 1966 he agreed to an amicable merger with the British Motor Corporation (BMC).

At a stroke this provided Jaguar with independent financial backing (if it was ever needed), and it brought Jaguar and its body shell supplier, Pressed Steel, into the same net. Although all parties stressed that this was a joining of forces rather than a take-over, the fact that BMC was making 20,000 vehicles a week, by comparison with Jaguar's 600, told its own story. Jaguar, at least, was guaranteed its independence of operation.

## XJ6 ON SALE – THE FIRST MONTHS

In several ways, the Jaguar 420 of 1966–8 did a great, though short-lived, job in pre-paring the public for the style of the new XJ6. We now know, in fact, that the XJ6 style had been finalized *before* that of the 420 – but that the 420 appeared two years ahead of it!

After he had retired, Bob Knight recalled that the XJ6 style had been finalized in 1964–5, but that Sir William Lyons then realized that it would take at least three years to get an entirely new shell tooled, the running gear developed, and the ensemble made ready for sale:

settings, the steering response, the rubber mounting settings, and the general hand-ling balance. The result was a large car which felt small, an executive saloon which felt like a sports saloon, and a high-performance car with all the silky and silent good manners of a town carriage. When the new car was revealed in September 1968, Jaguar's opposition had a real shock.

All the critical development on this new

*The XJ6 had wide-opening doors, with trim and stowage panels neatly detailed.*

*Jaguar – the gold badge on the nose of the first XJ6s. Having placed this symbol up front, it wasn't necessary to add any more decoration.*

*Perhaps those small extra grilles weren't needed to channel air into the engine bay or the front suspension area, but they added interest to the front corner of the car. Like the Mk X and 420 models before it, the XJ6 had four headlamps, two 7in and two 5¾in Lucas units.*

For the XJ6, Jaguar provided two fuel tanks, one in each corner of the boot, and twin fillers, with neatly detailed snap action caps on top of the rear wings.

The majority of early type XJ6s were built with automatic transmission, and that proportion increased further in the 1970s.

The XK engine was always an impressive and purposeful-looking power unit. In its XJ6 guise (this is a 2.8-litre example) it had ribbed cam covers, and all the accessories were neatly packaged.

... the XJ6 was not going to be announced soon enough, so then came the saga of the 420. In late 1965, Sir William became convinced that a further modification to the S-type was required ... So he reworked the front end and went to Pressed Steel stating that he wanted it in production by next July, even though at that stage there were no tools, or even drawings ...

In the end, although Pressed Steel bosses had originally said that the proposed timing was impossible, Sir William got his way (he usually did!), a new front end was prepared, and the 420 went on sale in October 1966.

There was a further complication. Even while the XJ6 was being prepared for production, the financial scene changed again. Sir William Lyons, who had become a director of British Motor Holdings after the merger of 1966, saw BMH merge with Leyland Motors in January 1968. British Leyland had a much more centralized policy strategy than BMH had ever had, so although there was no influence on Jaguar's fortunes in the short term, this would become more and more obvious in the years that followed.

But if Jaguar's directors had been worried about the reception the XJ6 would receive, they were soon reassured. The new car made a triumphant appearance in September 1968, with banner headlines in the motoring press, and with a glowing reception from the national and international media. It was, of course, the first all-new Jaguar shape to have been unveiled since 1961, and every enthusiast was delighted to welcome it.

In its issue of 26 September 1968, Britain's *Autocar* magazine summed up the general reaction like this:

Without a doubt the new Jaguar XJ6 model announced today is the most important new British car of 1968 and will be the centre of attraction at the Earls Court Show next month.

Series production of the new cars began in September 1968, so that the first UK deliveries could follow before the end of the year. At first there were three different derivatives, equipped and marketed as follows:

| | |
|---|---|
| XJ6 2.8-litre Standard | £1,797 |
| XJ6 2.8-litre De Luxe | £1,897 |
| XJ6 4.2-litre | £2,253 |

Overdrive was an optional extra at £61.33 on all types, while Borg Warner automatic transmission cost £102.22 on 2.8-litre types, and £144.39 on the 4.2-litre.

In pure marketing terms the 2.8-litre Standard model was something of a 'loss-leader'. It cost £100 less than the De Luxe, but had Ambla instead of leather upholstery, no rear seat centre arm-rest, no rear seat heating, the deletion of several other trim and furnishing details, and no power-assisted steering, though this was an optional extra. At first glance the ultra-low price was appealing, but Jaguar was not too distressed when its customers unerringly chose the De Luxe model instead. As you might guess, very few of these Standard cars were ever sold, and the model was officially dropped in March 1972.

Even though Jaguar did not make any road test cars available until 1969, queues to buy the cars built up very rapidly, and soon there were long waiting lists, with premium prices being achieved for 'young' second-hand cars.

Between the summer of 1968 and the end of 1969, there was a radical change of scene on the assembly lines at Browns Lane, and Jaguar's rationalization plan finally began to make sense. Before the first XJ6s were assembled, there were several lines building 240/340 types, S-types, the 420/Daimler Sovereign models, the Daimler V8-250s, the massive 420Gs, and of course the charismatic E-types – five different shapes in one overcrowded factory.

The weeding-out followed at once. Even before XJ6 build began, the S-types had disappeared and the 420/Daimler Sovereign models were then discontinued (though the Sovereign was listed for several months). By the spring of 1969, the 240/340 had been swept away, along with the Daimler V8–250.

Next it was time to start building the Daimler Sovereign version of the XJ6, but as the only technical difference was in standardizing the overdrive, and the only decorative changes were to the front grille and the decoration, this was no hardship.

From October 1969, therefore, sanity had returned to Browns Lane, which was producing a range of XJ6-based models, the 420Gs, and the E-types. From June 1970, in fact, the 420G model was also dropped, which left Browns Lane producing a multitude of XJ6s, and E-types. It was no wonder that production

forged ahead – from 22,650 cars in 1967 to 32,589 in 1972.

## 1969–1973: MEETING THE DEMAND

Once the XJ6 reached the showrooms, Jaguar's problem was not to sell enough cars, but to build enough to satisfy demand. Priority was given to sales in overseas markets, most particularly to North America, the result being that there never seemed to be enough supplies for the home market.

Jaguar didn't need to do much advertising at this time, for the XJ6 seemed to generate its own publicity. Britain's *Car* magazine, which was not known for being over-sympathetic to British cars, made the XJ6 its 'Car of the Year', while Don, the brake linings firm, gave it the Don Safety Award.

When *Autocar* tested an automatic transmission 4.2-litre car in June 1969, it described the car as being 'Unbelievable value. The best there is', and ended by writing:

We of *Autocar* set it as a new yardstick, a tremendous advance guaranteed to put it ahead for several years at least.

When Sir William Lyons read the Lord Wakefield Gold Medal Paper to the IMI in April 1969, he made this very telling point:

By the judicious use of the right type of promotion ... the specialist manufacturer must endow his products with that aura of 'exclusiveness' which is the hallmark of the true specialist car. A manufacturer cannot be far off the mark if it is the ambition of every keen motorist to own one of his cars. Such a situation exists with the XJ6.

And so it did. At the end of 1969, readers of *Modern Motoring* of Australia voted the XJ6 as the 'best all-round car regardless of cost', and by 1970 there was so much pent-up frustration over deliveries that a group of Swiss customers actually protested about the long wait for deliveries outside British Leyland's Berkeley Square offices in London.

Jaguar did its best to meet the demands. The 50,000th XJ6/Sovereign car was built in the summer of 1971, by which time no fewer than 28,000 had been exported. At that time more than 650 XJ6/Sovereign cars were being produced every week, of which more than 100 were being sent to the USA.

Unhappily, the business was gradually being infected by the strike-prone habits of other British Leyland workers, and a ten-week stoppage in mid-1972 did not exactly boost the company's reputation. Even so, Lord Stokes, Chairman of British Leyland, stated flatly that 'Jaguar is the top of the British Leyland range and is going to stay there . . . Jaguar needs British Leyland, but clearly British Leyland needs Jaguar and we intend to develop the company to its utmost.'

By this time Jaguar was reducing the waiting lists for its new car, and an increasing number of customers were revelling in the spacious refinement which an XJ6 could give them. The cars had built up a fine reputation, except for the niggling doubts about the build quality and – in North America (where Lucas was known as 'The Prince of Darkness') – the performance of the electrical systems. During this time, too, Jaguar had been working hard on the cars' specifications, on improvements to the design, and was keeping a wary eye on legislative developments in the USA. The first major face-lift to the design – the Series II – was on the way.

# 3 XJ12 – The World's Most Refined Car?

This is no time for platitudes, or half-truths. The fact is that, when announced, the Jaguar XJ12 was an extraordinary car. By combining a phenomenal new V12 engine with the good looks and behaviour of the existing XJ6 structure, Jaguar produced a superlative piece of engineering. The more cynical motoring writers – a group which, at the time, was burgeoning apace – complained about the steering feel, the fuel consumption, the build quality, or whatever else they thought should be stirred up to make controversy, but the facts could not be denied. Here was a car so refined, so fast, yet so flexible that it was in a class of its own. To summarize I need only quote *Autocar*, from its road test of March 1973 which, after describing it as a 'truly outstanding car', summed it up as '. . . a car of great superlatives. It is a marvellous achievement, deservedly the envy of the world.'

The combination of a V12 engine with the XJ6 structure had been foreseen from the very start of the new car's development, though finalization of the V12 engine was much delayed. For Jaguar enthusiasts, as well as for Jaguar itself, this was deeply frustrating. As early as spring 1968, when privileged motoring writers were being briefed about the forthcoming new models, Jaguar spokesmen were happy to mention V12- *and* V8-engined derivatives, but as the launch passed, and as waiting lists for the six-cylinder engined XJ6 continued to grow, there was no further news of new-generation engines. Not only was Jaguar determined to get the V12 design absolutely

right before putting it on sale, it also had to make the very serious decision to spend

---

**Stillborn V8**

When Jaguar decided to make a production engine out of the V12 engine in the 1960s, Walter Hassan and Harry Mundy were also asked to design a V8 version of it. In this way Sir William Lyons hoped to be able to offer a choice of straight-six, V8 and V12 engines in the same graceful bodyshell.

The V8 was little more than a chopped-off version of the V12, which is to say that it had a 60-degree vee angle, and the same single-overhead-camshaft valve gear. In standard form it would have had a capacity of 3,562cc – at a time when the six could easily have been a 3.0-litre unit.

Unhappily, the V8 engine was not a success. As Walter Hassan later wrote in his autobiography *Climax in Coventry* 'the V8 . . . was a disappointment to us all. Engineers know that V8s should have 90 degrees between the banks, but we had hoped that the unpleasant secondary vibrations involved in a 60-degree design could be suppressed by clever engine mounting. But this proved not to be possible and, as with the four-cylinder XK engine of the 1940s, there were unpleasant vibrations in the structure, felt also through the gear lever, which we could not tame.'

Other people commented that the V8-engined car somehow 'didn't feel like a Jaguar', so the project was abandoned. Jaguar must still be sensitive about this design, for I have never seen an engine, nor even a picture of an engine.

huge amounts on tooling to build engines in quantity at the Radford works. One must remember that this was probably the most complex quantity-production engine ever produced in the UK. In the event, the V8 alternative was cancelled, the V12 engine was unveiled in 1971, for use in the E-type Series III, and the introduction of the XJ12 was delayed until the summer of 1972.

## V12 ORIGINS

When I was a junior designer at Jaguar in the late 1950s, the first XK-derived 60-degree V12 engine had already been schemed out, though an engine had not then been built. At this time it was purely meant to be a racing engine for use in long-distance sports car events. Jaguar saw that the XK engine was already reaching its limits, and needed a more powerful engine with which to beat the might of Ferrari, Maserati and Mercedes-Benz.

The choice of a V12 layout was natural, so logical in everyone's eyes that no other alternative seems to have been considered. By choosing a four-cam V12 Jaguar could make a large engine which used many proven XK engine components, including valve gear and camshafts, though the cylinder heads, complete with 'downdraught' inlet ports, were completely different. By choosing a V12, too, Jaguar could match Ferrari head on, size for size, and capacity for capacity!

If Jaguar had not withdrawn from motor racing at the end of 1956, and *if* it had come back with a new car to succeed the D-type, the racing V12 engine might have been finalized and developed by the late 1950s, but in the event the design of the engine *and* the car which was to use it were much delayed. Although Jaguar began to dabble with the design of a new two-seater racing car, coded XJ13, in 1960, the very first Jaguar V12 engine, a 4,994cc unit, did not even run on a

### The Coventry-Climax Connection

By the time Jaguar bought Coventry-Climax, in 1963, it was a world-famous concern, not only in the field of motor racing, but as a manufacturer of industrial and stationary engines.

The Coventry-Climax company was set up by H. Pelham Lee in 1917, as a successor to a company which had built Coventry Simplex engines. Between the wars 'The Climax' not only built a series of fine engines – notably used by Triumph in the 1930s – but also developed fire-pump engines, which were omnipresent in World War II.

Leonard Lee, H. Pelham Lee's son, attracted Walter Hassan, with all his early XK engine experience, to Coventry-Climax in 1950. Along with Harry Mundy, Walter then inspired the birth of many fine racing engines, and new fire-pump designs, in the next thirteen years.

Leonard Lee agreed to sell his company to Jaguar in 1963, though little integration took place until both concerns found themselves in the British Leyland Group. Walter Hassan, however, began consulting with his Jaguar colleagues almost at once, persuaded Harry Mundy to join Jaguar in 1964, and eventually moved across, full time, to Jaguar in 1966.

Sir William Lyons was more of a pragmatic businessman than Leonard Lee, who loved racing and loved to have new designs all around him. This explains why the Coventry-Climax racing programme was allowed to reach its peak in the 1.5-litre Grand Prix arena of 1961–1965, but why no engine was ever designed for the new 3-litre formula which came into force in 1966.

Once British Leyland had been nationalized, and companies began to be closed down or hived off, there was no place for Coventry-Climax in the scheme of things, so it was sold off once more. There is now little surviving evidence of this illustrious concern in the Coventry area.

test bed until August 1964. By the time it was ready to be used in a race car, it was fitted with Lucas fuel injection and developed

502bhp at 7,600rpm. One might rightly accuse Jaguar of not being serious about a return to motorsport at this stage, for building work on the XJ13 car was not even started until mid-1965, after Ford's massively powerful GTs had already competed twice at Le Mans, and although that car was complete in March 1966 it did not make its first test run until March 1967. However, although the XJ13 lapped MIRA's outer circuit at 161.6mph (260kph), and set competitive times in brief tests carried out at Silverstone, it was clear that a lot of work would have been needed to make it into a winner. Regretfully, therefore, XJ13 was shelved, and did not stir again until 1971 – when it was wrecked while meeting a photo-publicity engagement at

MIRA ahead of the launch of the *production* V12 engine.

## DEVELOPING THE PRODUCTION ENGINE

In the meantime, Jaguar had taken control of Coventry-Climax and, coincidentally, had also decided to turn the V12 into a road-car engine. To take charge of all engine development work, former Coventry-Climax designer Harry Mundy joined Jaguar, Claude Baily carried on his work as Chief Engine Designer for a time, and (from 1966) Walter Hassan moved across from Coventry-Climax to succeed him.

---

### Walter Hassan (born 1905)

The man who started his working life with Bentley in 1920, and who had still not ended it in the early 1990s, had three spells at SS Cars (later Jaguar Cars) which were bracketed by an equally illustrious period with Coventry-Climax. Not only was he famous as an engine designer in post-war years, but as a racing mechanic in the early days, and as a consultant engineer to F1 teams for years after his retirement.

Walter Thomas Frederick Hassan was born in London in 1905, attended the Northern Polytechnic in Holloway and the Hackney Institute of Engineering, but left the latter establishment on his fifteenth birthday. He joined Bentley as an apprentice, becoming that company's fourteenth employee, and stayed at Cricklewood until Bentley went into receivership in 1931.

After becoming a racing mechanic he moved on to be ex-Bentley Chairman Woolf Barnato's personal race car engineer, designed and built the Barnato-Hassan and Pacey-Hassan Brooklands 'specials', spent time with ERA in Bourne, then at Thomson & Taylor at Brooklands in the 1930s, before moving to Coventry to join SS Cars as Chief Development Engineer in 1938.

Following a brief spell with the Bristol Aeroplane Company during the War, he returned to SS in 1943. After dabbling with the design of lightweight parachute-drop fighting vehicles, he next got involved in the design of the post-war XF/XG/XJ/XK engine family, and ran the company's busy experimental department.

In 1950 he became Coventry-Climax's Chief Engineer, later Technical Director, but after Coventy-Climax was taken over by Jaguar in 1963 he found himself working on two sites in tandem – Widdrington Road (Coventry-Climax) and Browns Lane (Jaguar).

He officially returned full time to Jaguar as Group Chief Director in 1966, took over from William Heynes as Engineering Director in 1969, and finally retired in 1972. During that time Jaguar introduced the XJ6 family, the V12 engine, and the first two cars which accepted it – the Series III E-type and the XJ12.

He still found himself in demand as a consultant, spent some time trying to make sense of the BRM V12 F1 engine, then continued to dabble in engine design work until he was well into his eighties.

The story of the transformation of the V12 engine, from a peaky 'racer' with four overhead camshafts, to a silky and refined two-cam road-car unit, is so well known that I need only summarize it. Having seen the result of race-engine development work at Coventry-Climax, Hassan and Mundy were not surprised to discover that the new Jaguar V12 had both a lack of top end power *and* mid-range torque. This was eventually blamed on the original 'downdraught' inlet port arrangement, which was by no means as efficient as hoped. A series of different single-cylinder test engines were then produced, to test a variety of cylinder head and valve gear arrangements. Work was then

*When the XJ12/Double-Six was launched in 1972, many pundits gave it the title of the 'world's most refined car', which did nothing for the blood-pressure of Rolls-Royce, Cadillac or Mercedes-Benz bosses. From the front there was very little to signal the use of a V12 engine – only the grille itself was different from that of the XJ6/Sovereign. This particular car is a Daimler Double-Six.*

*If you ever got this close to the tail of the V12-engined saloon, you could just pick out the badge which told the story – in this case 'Double-Six' to the right of the number plate.*

concentrated on an alternative layout, with vertical valves, a single-overhead-cam lay-out, a flat cylinder head, and – for a time – it was thought that *both* types of head should be put into production. Back to back tests of such V12s – both on the test bed, and in development cars – then showed the 'flat-head'/single-cam arrangement to be better in all respects so, after a great deal of heart-searching, this was chosen for production. When tested in cars the flat-head single-cam arrangement gave more low-speed torque and 'slogging power' *and* it also accelerated better at higher engine speeds.

There was no longer any doubt in anyone's

minds. Jaguar therefore 'pressed the button' on tooling layouts in 1968, began the expensive business of installing new machinery at the ex-Daimler Radford factory, and prepared to build V12-engined production cars in the early 1970s.

Other than in its general layout, and its 60-degree vee angle, the series-production V12 differed in almost every way from the original racing type. The difference in heads has already been explained, but the cylinder block was now to be of 'open-deck' construction, the sides of the block wrapped around the crankshaft, and the bore had been opened up to 90mm, which meant that the

## Harry Mundy (1914–1988)

The design of the Jaguar V12 engine, in conjunction with Walter Hassan, was the high point of Harry Mundy's impressive career, not only as an engineer, but as a journalist.

Born in Coventry, educated in the city, then apprenticed to Alvis, he joined ERA in 1936 and spent a short time with Morris Engines (also in Coventry) before joining the RAF as an Engineer Officer during World War II.

After the war, he joined the design team at BRM, working on the controversial V16 Grand Prix car, but in 1950 Walter Hassan persuaded him to join him at Coventry-Climax, to be his Chief Designer.

In five short years at Coventry-Climax, Harry not only laid out the FWA sports car engine, the stillborn 2.5-litre FPE V8 F1 engine, and the famous four-cylinder FPF twin-cam racing engine, but was involved in other industrial engines.

He joined *The Autocar* as Technical Editor in 1955, and became a somewhat tempestuous journalist for nine years, during which he also schemed up the original Lotus-Ford twin-cam engine. His irascible, irrepressible character, soon became a legend in the motor industry.

Walter Hassan then attracted Harry back to Coventry, where he became Jaguar's Executive Director of Power Unit Design, and headed the design team which developed the V12 engine. Later, he also initiated design of the new-generation AJ6 family of six-cylinder engines, which were launched after he had retired.

He retired from Jaguar in 1980, and died in 1988, aged 72.

came to fit it to cars, the engineers had to exercise a great deal of ingenuity to cram everything into place.

At this point, Jaguar also abandoned the racing-type Lucas fuel injection systems and opted for multi-carburettor installations instead. Engine design chief Walter Hassan always regretted that the use of injection had to be postponed, but after he had retired it was standardized in 1975.

The first Jaguar saloon to be fitted with the V12 engine was a Mark X development car, but definitive engines were running in E-types and XJ6 models by the end of the 1960s. Even so, the launch of production cars, once forecast for 1970, had to be delayed until 1972.

## XJ12 – JAGUAR'S NEW FLAGSHIP

No sooner had the V12-engined E-type Series III sports car been launched in 1971, than the pundits began talking about a future V12-engined saloon. It was far easier for them to forecast what Jaguar would do, than for Jaguar to achieve its aims, but it was always quite obvious what the company's plans would be. The strategy was so logical, in fact, that pundits began quoting the correct model title – XJ12 – a long time before Jaguar admitted to its existence.

As early as 1968, the company had stated that 'V-engines will be offered as options on the XJ model within the next two years', so it was clear that the E-type's V12 was also destined for use in the XJ6 shell. When the time came, however, the transplant was neither simple, nor easy, and a large number of other changes were made at the same time, to create an integrated package.

But it wasn't easy. It is true that in 1964 Jaguar had considered alternative engines by designing a spacious engine bay for its new car, but in 1964 the V12 engine was still at an early stage of its development.

swept volume was to be 5,343cc. Not only that, but the new single-cam head was simpler to cast and machine, and each complete assembly was no less than 161lb (7.26kg) lighter than the twin-cam. It was a smaller, lower and altogether more compact engine – which was fortuitous because when Jaguar

The fact was that installing the production two-cam V12 was very difficult – it was a classic 'shoe-horn job' – and that the four-cam V12 was probably too bulky ever to have been slotted into place! You only have to open up the bonnet of a V12-engined Jaguar saloon to see why there was a mechanic's joke circulating at the time. Called the 'cup of tea test', the story went that after the engine was installed, and all the auxiliaries were hooked up, a cup of tea was poured in to the top of the engine bay. If any of the tea dripped through to the ground, then something had not been fitted!

Other than a different style of radiator grille for the new XJ12 – it had vertical slats rather than a venetian blind theme – there were no changes to the front-end style. This meant that no extra cooling air could be channelled into the radiator, and into the engine bay. On the E-type there were louvres in the bonnet panel, but this palliative was not chosen for the saloons.

This could have been a serious problem, but Bob Knight's development engineers solved it in a unique manner. The 60-degree V12 engine, complete with four Zenith-Stromberg carburettors, their air cleaners and associated trunking, was an impressive package, and by the time other ancillary equipment (including the air-conditioning which was essential for sale to American customers) had been fitted, the engine bay was full, and likely to become very hot. In fact, the only major problem was in, of all things, keeping the battery cool. This was achieved by providing the battery with its own container, into which a special thermostatically controlled cooling fan was placed, fed with fresh air through flexible trunking. If the customer, in other words, was to have a V12 engine, he would have to put up with the complications!

Although the general layout of the existing XJ6 was retained, there had been time to look at every corner of the car, to match up to the silky new 265bhp engine.

The most important decision regarding the running gear was that there was to be no manual transmission option for the new V12-engined car. Jaguar's existing four-speed all-synchromesh gearbox, when matched to this installation, was not really man enough for the job (it had an easier time in the E-type Series III, which was lighter and not nearly as refined), so the only transmission on offer was the Borg Warner Model 12 three-speed automatic. In later years, the irrepressible Harry Mundy (who had started his working life designing gearboxes at Alvis in the 1930s) found time to design a new five-speed manual gearbox to suit this engine, but it never went into production.

Because the new aluminium V12 only weighed 80lb (36kg) more than the iron-blocked 4.2-litre 'six', the entire car weighed just 123lb (56kg) more than the XJ6, so except for the use of stiffer front springs, the suspension and handling balance needed little attention. Ventilated front discs were standardized, along with a brake balancing valve in the rear circuit, and there was a new type of 70-section Dunlop radial ply tyres, with a steel breaker strip and a nylon casing.

Visually, the XJ12 was an unassuming supercar. Apart from the new front grille, and the XJ12 badges on the boot lid, it was virtually indistinguishable from the XJ6, while inside the car there were new door trim panels. The *cognoscenti*, of course, could listen for the slightly different exhaust note, for both cars were equipped with twin exhaust pipes . . .

Like any other Jaguar of the period, the XJ12 looked, and was, a real automotive bargain, for it offered a top speed of 150mph (241kph), yet was priced at a mere £3,726. It was, no question, the world's fastest saloon car, and there were many prepared to swear that it was the world's most refined car as well.

Almost at once, stories of the engine's legendary smoothness began to circulate. In

*The V12-engined saloons were, and are, extremely desirable and very fast machines, with a top speed of 150mph (240kph) which was reached completely without fuss or strain.*

particular, it was possible to sit inside the car, and not be able to decide if the engine was running at idle speed, or was switched off. Even a blip of the throttle foot produced no more than a tiny tremor inside the cabin; in fact, the best way to be sure was to look at the 7,000rpm rev counter. It was, if anything, an even smoother, more poised, and more accomplished car than the six-cylinder XJ6, and in view of the relatively low price asked, this was a quite astonishing achievement. Engineers at Rolls-Royce, Cadillac and Mercedes-Benz, in particular, made haste to buy an XJ12 to study how Jaguar

had achieved this. This, incidentally, was at a time when the automatic transmission version of the XJ6 cost £3,229 – and when the Mercedes-Benz 300SEL 6.3-litre V8, which was genuine competition, was priced at £8,600!

## XJ12 ON THE MARKET – A LONG-RUNNING SAGA

At the time of writing, in 1991, the V12-engined Jaguar saloon has been on the market for nineteen years, and still sells steadily at the rate of about 1,500 cars a year. That's not many, you might say – and you would be right – but we must remember that the body style was outdated in 1986, when the all-new six-cylinder XJ40 models were launched, and that a marriage between that shell and the V12 engine is confidently expected in the early 1990s.

When Jaguar pencilled in a launch date of July 1972, it was expecting to start delivering cars at once to UK customers, and to 'fill the pipeline' to the USA so that the new cars could then go on sale as 1973 models. All very commendable, of course – if the workforce had not chosen to go on strike at the same time! The result was that the XJ12 was introduced when the Browns Lane production lines had already been at a standstill for four weeks; the strike (over pay and conditions) dragged on for a further six weeks, which meant that the first XJ12s did not reach their customers until October 1972.

*From the side view, nothing tells the casual observer (or the traffic policeman!) that this is a V12-engined car.*

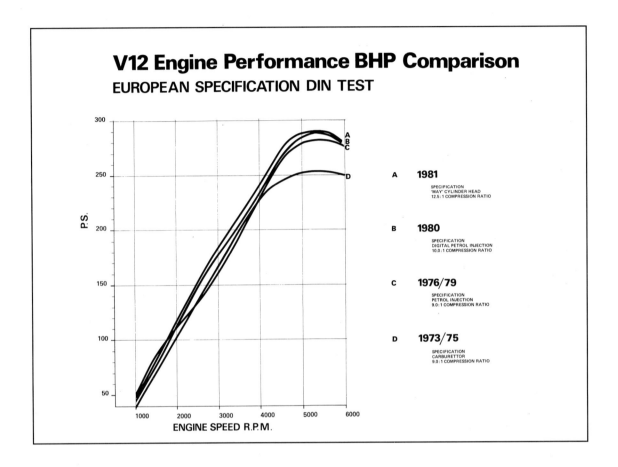

# V12 Engine Performance BHP Comparison
## EUROPEAN SPECIFICATION DIN TEST

**A  1981**
SPECIFICATION
'MAY' CYLINDER HEAD
12.5:1 COMPRESSION RATIO

**B  1980**
SPECIFICATION
DIGITAL PETROL INJECTION
10.0:1 COMPRESSION RATIO

**C  1976/79**
SPECIFICATION
PETROL INJECTION
9.0:1 COMPRESSION RATIO

**D  1973/75**
SPECIFICATION
CARBURETTOR
9.0:1 COMPRESSION RATIO

*These two sets of curves show how the power and torque of the 5.3-litre V12 changed significantly between the early 1970s, when it was equipped with four carburettors, and 1981, when the fuel-injected unit was completely redeveloped in 'May' cylinder head form. During that time, peak power increased by about 40bhp, and peak torque was lifted from 296lb ft to nearly 320lb ft.*

In the meantime, Jaguar had also launched the XJ12's sister car, which it chose to call Daimler Double-Six. The choice of 'Double-Six' for the Daimler-badged car was a stroke of genius on someone's part (was it Lofty England, who had been a Daimler apprentice between 1927 and 1932?), for it harked back to the great days of the 1920s and 1930s, when the independent Daimler company had produced a series of V12-engined models; in those days, however, the engines were low-revving sleeve valve units,

and the chassis were usually bodied as limousines with top speeds of around 80–85mph (129–137kph). Times had changed!

The Daimler Double-Six was to all intents and purposes the same car as the XJ12, except for the badging and a few different items of trim. It was also slightly more expensive – £3,848 against £3,726. When the twin V12-engined saloons went on sale, Jaguar revealed that its Radford factory could produce up to 200 engines every week – around 10,000 a year – half of which were

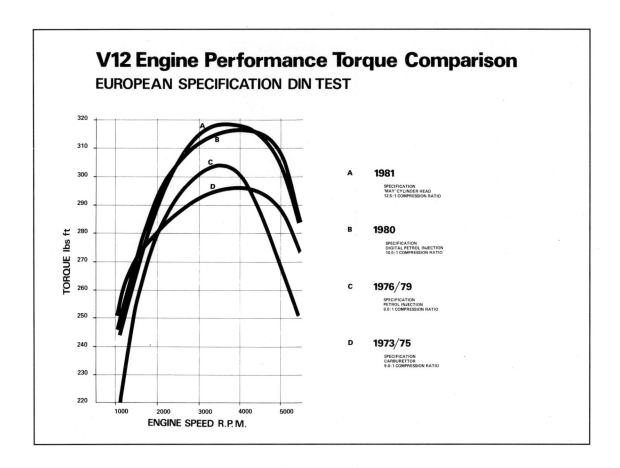

## V12 Engine Performance Torque Comparison
### EUROPEAN SPECIFICATION DIN TEST

**A**   **1981**
SPECIFICATION
'MAY' CYLINDER HEAD
12.5:1 COMPRESSION RATIO

**B**   **1980**
SPECIFICATION
DIGITAL PETROL INJECTION
10.0:1 COMPRESSION RATIO

**C**   **1976/79**
SPECIFICATION
PETROL INJECTION
9.0:1 COMPRESSION RATIO

**D**   **1973/75**
SPECIFICATION
CARBURETTOR
9.0:1 COMPRESSION RATIO

earmarked for the Series III E-type roadster. At the time, before the Yom Kippur war erupted in the Middle East, and before the energy crisis whipped up crude oil prices all around the world, Jaguar seemed to be capable of using every one of these engines.

Even before the end of 1972, Jaguar had expanded the range of V12 engines, to four slightly different models. One reason was the launch of the longer-wheelbase version of the body shell, in which an extra 4in (10cm) was let into the floorpan, roof and rear doors, behind the line of the front seats, the other to the use of the 'Vanden Plas' coachbuilding expertise for an extra-luxurious trim pack on the Daimler Double-Six. By the end of the year, therefore, the British customer could buy these XJ12-based cars:

| | |
|---|---|
| XJ12 (normal wheelbase) | £3,726 |
| XJ12L (long wheelbase) | £4,052 |
| Daimler Double-Six (normal wheelbase) | £3,849 |
| Daimler Vanden Plas (longer wheelbase) | £5,439 |

When the Vanden Plas was introduced, *Autocar* rather waspishly pointed out that 'VDP motoring cost £400 an inch more than the standard car for the extra four inches in the wheelbase.'

What did the Vanden Plas customer get for his extra £1,590, or 41 per cent? Apart from the extra rear-seat leg room, and the trimming and painting at the Vanden Plas coachbuilding works in North-West London, the brutal answer would have to be – not a lot.

*The only difference in the front style of the V12-engined car was a different arrangement of the grille bars.*

*Early examples of the V12 engine were equipped with four carburettors – fuel injection did not appear until 1975. The nightmare complication of wire, pipes and linkages would gradually be tidied up over the years.*

*The fascia layout of the early 1970s V12 saloons (this is a Daimler model) was packed with instruments and switches. One look at the calibration of the rev-counter and speedometer of this car tells you that it means business.*

*The miracle is that enough cooling air ever got past the narrow bars of the V12-engined car's grille. There's no doubt that the engine bay could get very hot – the battery even had its own individual cooling fan.*

*Double-Six not only defined the layout of the V12 Daimler's engine, but it revived a famous Daimler name of the 1920s and 1930s, when different generations of V12 were fitted.*

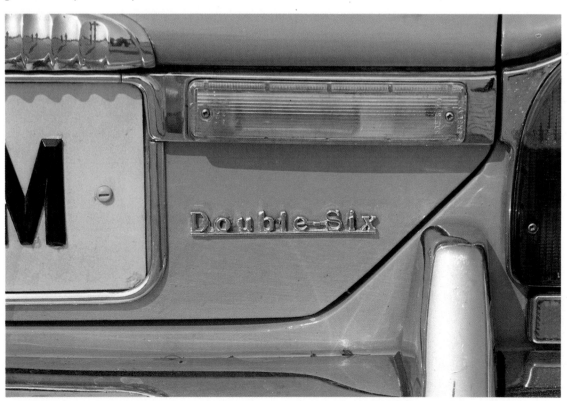

## XJ12 and XJ12L Series I (1972 and 1973)

### Layout
Unit-construction steel monocoque five-seater, front engine/rear-drive, sold as four-door saloon

### Engine

| | |
|---|---|
| Block material | Cast aluminium |
| Head material | Cast aluminium |
| Cylinders | 12 in 60-degree vee |
| Cooling | Water |
| Bore and stroke | 90 × 70mm |
| Capacity | 5,343cc |
| Main bearings | 7 |
| Valves | 2 per cylinder; single OHC operation |
| Compression ratio | 9.0:1 (8.0:1 optional) |
| Carburettors | 4 Zenith-Stromberg 175CD |
| Max. power (DIN) | 265bhp @ 5,850rpm |
| Max. torque | 304lb/ft @ 3,500rpm |

### Transmission
Manual transmission not available. Borg Warner Model 12 automatic transmission, with torque converter

### Internal ratios

| | |
|---|---|
| Top | 1.00:1 |
| 2nd | 1.45:1 |
| 1st | 2.40:1 |
| Reverse | 2.00:1 |
| Maximum converter multiplication | 2.0:1 |
| Final drive | 3.31:1 |

### Suspension steering and brakes
All as for contemporary six-cylinder XJ6 models

### Dimensions
As for contemporary six-cylinder XJ6 models, except for:

| | |
|---|---|
| Unladen weight | 3,881lb/1,760kg (shorter wheelbase) |
| | 4,116lb/1,879kg (longer wheelbase) |

### XJ12 Series III (1973 to 1975) XJ5.3 (1975 to 1979)

Specification as Series I type except for:

### Engine
As Series I at first, then from Spring 1975 (XJ5.3):

| | |
|---|---|
| Fuel injection | Bosch-Lucas |
| Max. power (DIN) | 285bhp @ 5,750rpm |
| Max. torque | 294lb/ft @ 3,500rpm |

### Transmission
From April 1977:

GM400 automatic transmission, with torque converter

## Internal ratios

| | |
|---|---|
| Top | 1.00:1 |
| 2nd | 1.48:1 |
| 1st | 2.48:1 |
| Reverse | 2.08:1 |
| Maximum torque multiplication | 2.4:1 |
| Final drive | 3.07:1 or 3.31:1 |

### XJ12C/XJ5.3C (1973 to 1977)

Specification as for contemporary Series II XJ12 types except for:

## Layout

Sold only on short wheelbase, as two-door saloon (C = Coupé, on badge). Four-door version not available

## Dimensions (in/mm)

| | |
|---|---|
| Wheelbase | 108.8/2,763 |
| Overall length | 189.5/4,813 |
| Unladen weight | 3,885lb/1,760kg |

### XJ5.3/Sovereign V12 Series III (Introduced 1979)

Specification as for late-model XJ12 Series II except for:

## Engine (From July 1981)

| | |
|---|---|
| Compression ratio | 12.5:1 |
| Max. power (DIN) | 299bhp @ 5,500rpm |
| Max. torque | 318lb/ft @ 3,000rpm |

Note: By 1990 there was a catalyst version of the engine:

| | |
|---|---|
| Compression ratio | 11.5:1 |
| Max. power (DIN) | 264bhp @ 5,250rpm |
| Max. torque | 278lb/ft @ 2,750rpm |

## Transmission

| | |
|---|---|
| Final drive ratio | 2.88:1 |

## Suspension, steering and brakes

As contemporary six-cylinder XJ6 models, with cast aluminium wheels

## Dimensions

As for Series III Jaguar XJ6 models, except for:

| | |
|---|---|
| Unladen weight | 4,234lb/1,920kg |

Mechanically, of course, there were no differences, but the seats were special, all windows had electric lifts, there was air-conditioning as standard, and there was Sundym glass all round. Even so, the Daimler options pack which had been standardized was only worth £469 on the Daimler list, the rest being put down to exclusivity.

A year later Jaguar introduced the Series II models, (described in detail in the next chapter), which resulted in the V12-engined cars picking up the same changes and improvements. In addition, V12-engined/automatic transmission versions of the brand-new normal-wheelbase two-door Coupé were also previewed, though these did not actually go on sale until the spring of 1975; this fascinating, but ultimately short-lived, model, is also described in detail in the next chapter.

Almost at once, Jaguar (and any number of other companies that built large-engined and thirsty cars) was hit badly by the onset of the Yom Kippur war in the Middle East, and by the energy crisis and the rocketing of petrol prices which followed it. Within weeks, the worldwide demand for cars like the XJ12, which could usually only return 12–13mpg (23.6–21.8l/100km), slumped alarmingly, and the future of the V12 engine was thrown into doubt.

Within a year, however, the energy crisis had receded (though petrol prices did not come back to pre-war levels), Jaguar had dropped the original shorter-wheelbase version of the four-door body shell, and it was preparing to improve the V12 engine's efficiency. The result was unveiled in May 1975, when all the 12-cylinder engines were given Bosch-Lucas electronic fuel injection. The outcome was not only a more economical and fuel-efficient unit, but one which was significantly more powerful than before. Here is a direct comparison between the two types:

1 *V12 with carbs*: 253bhp (DIN) at 6,000rpm; 302lb ft torque at 3,500rpm
2 *V12 with fuel injection*: 285bhp (DIN) at 5,750rpm; 294lb ft torque at 3,500rpm

The fuel injection, therefore, had helped boost peak power by no less than 12.6 per cent. At the same time the final drive ratio was 'lengthened' from 3.31:1 to 3.07:1, the overall result being that average fuel consumption had improved from about 12mpg (23.6 l/100km) to about 13.5mpg (21 l/100km). It wasn't much, but it was a definite step in the right direction.

The V12-engined two-door Coupés – XJ5.3C, and Daimler Double-Six two-door – were fast and desirable, but sold very slowly. In two and a half years, only 1,855 Jaguar-badged types, and a mere 407 Daimlers were produced. The high-profile and very public failure of the Broadspeed-prepared race cars in 1976 and 1977 cannot have helped their cause, but one still senses the relief with which Jaguar closed *that* chapter on the XJ6 story at the end of 1977.

By that time, Jaguar had updated the specification of the renamed (to XJ5.3) V12-engined cars yet again, for the old Borg Warner Model 12 automatic transmission was ditched in favour of the latest General Motors build GM400 automatic transmission in 1977. This was the transmission that Rolls-Royce had chosen for its latest Silver Shadows, and that was in use on all current Cadillac models, so Jaguar was in good company.

From 1979, the V12-engined cars progressed from Series II to Series III, picking up all the bodyshell, style and equipment improvements of the latest six-cylinder cars (*see* Chapter 4), though for the moment their engine specifications were left untouched. Then, in mid-1981, came 'HE', and a third generation of V12 engines. HE, quite simply, stood for 'High Efficiency', which summarized the changes that had been made to the 5.3-litre V12 engine. The pioneering work had been carried out by the Swiss engineer Michael May, the object having been to improve the breathing in the cylinder heads.

The new heads were officially called 'May heads' (Sir William Lyons would surely not have approved? In *his* day, there was no way that he would have given the names 'Ricardo' or 'Weslake' to any heads that were breathed upon by those experts . . . ) whose secret was to promote more swirl, better fuel-air mixing, and greater efficiency.

Externally, and visually, there was no change to the latest engines, which were still 5.3-litre units, but internally there were new pistons, different shapes to the ports and combustion chambers, and a dramatically higher compression ratio, of 12.5:1. The result was that peak power rose to 299bhp (DIN) at

*The V12 engine, here seen in early-1980s 'HE' form, was a massive, but beautifully detailed piece of engineering. It completely filled the engine bay of any car that used it, which made maintenance a long and costly business.*

5,000rpm, and 319lb ft of torque at 3,000rpm. As the final drive ratio was once again changed, to give ever-higher gearing (this time the ratio became 2.88:1), better fuel efficiency was expected.

This, then, was the definitive V12-engined Jaguar saloon, for in the next ten years few changes were to be made. It is true that there was a comprehensive reshuffle of model names for the 1984 season, which resulted in the Jaguar being renamed Sovereign HE, and in the Daimler Vanden Plas variety being dropped. More important, however, was the decision to keep on making the V12-engined Series III cars after the new-shape XJ40 model appeared in 1986. In fact, the XJ40 bodyshell was not originally intended to accept the V12 engine, which meant that

continuance of the 'old' Series III shell was inevitable until expensive changes were made.

For purely practical purposes, when the Browns Lane 'pilot plant' was freed of its function for early production of XJ40s, assembly of the Series III V12-engined cars was moved there from the main buildings alongside it. Even then, the innovations were not over for, after a long development period, the 1991 model was launched with ABS brakes as standard (these had previously not been available on the V12-engined saloons), and with a catalyst as standard.

By the early 1990s, however, the V12 engine had gradually slipped backwards in terms of efficiency, for in catalyst form it

*The V12-engined Series III models – Jaguar* (left) *and Daimler* (right) *were still selling steadily as the 1990s opened. A replacement, using the more modern 'XJ40' type of style, was under development by that time.*

only produced 264bhp, though independent road tests showed that 15–16mpg (18.8–17.7 l/100km) was now normal, which was at least 30 per cent better than it had been in the early 1970s.

By this time, of course, the car was ripe for replacement, and it suffered by direct comparison with contemporary V12-engined rivals from BMW and Mercedes-Benz. Yet its feline character, and its sheer elegance, was still unsurpassed, even at an asking price of £39,340 for the Jaguar, or £44,140

for the Daimler-badged car. Indeed, as *Autocar & Motor* staffmen wrote, in a comparison test published in July 1991 'And yet there's still something about the Jaguar – its innate restraint, good taste and grace – that the others lack entirely. It's a beautiful car and, for some, that will always be enough.'

By that time, total V12 saloon production was approaching 50,000 units, and it was clear that it *would* eventually be replaced. Didn't that make it a success by any standards?

# 4 Series II –
# Broadening the Range

Between 1973 and 1979, Jaguar reversed almost every policy it had laid down in the late 1960s. The object of launching the XJ6 had been to rationalize a complicated model range, and for a time this had been achieved. Then came the Series II cars – when Jaguar rediscovered what the magic world of product planning was all about.

In 1968 there had been one XJ6, with one wheelbase and one engine produced in two capacities. By 1976 this basic design had been varied so much that there had already been three different wheelbases, the use of six-cylinder and V12 engines, three different body styles – and in most cases a choice of Jaguar or Daimler badges. In part, this was due to pressure from British Leyland, who followed the North American ideal, and wanted to give their customers every possible choice of cars. In the event, though, was it any wonder that the customers became confused, that the Jaguar workforce sometimes found it difficult to produce cars of the appropriate quality, and that the Jaguar image was dented?

This period in the cars' history is so convoluted that to make it clear, I have had to write several different sections. The arrival of the V12 saloon models has already been covered (in Chapter 3), while the story of the XJ-S range follows in Chapter 5. I have also described the two-door Coupé models at the end of this chapter.

First of all, therefore, I should point out that the Series II models were in production from 1973 to 1979 – actually for about five and a half years – and that in almost every way they were very similar to the original (retrospectively known as Series I) models. When the Series II cars arrived, the choice of two wheelbases, two engine layouts, and two marque badges had already been established. As far as the saloons were concerned, these choices were retained during the life of the Series IIs. Along the way, however, yet another type of six-cylinder XK engine appeared and the V12 engine gained fuel injection. Jaguar also previewed a new type of manual transmission, but this was then held back until Series III build began in 1979.

## SERIES II –
## PROGRESSIVE
## IMPROVEMENTS

After five years, Jaguar decided that it was time to tidy up the range, to improve certain features, and to remind the public that this potentially great car could be made still better. This was also the ideal time to make some changes which would soon be made compulsory by new North American regulations. The basic styling was not altered, there being no reshaping of the bodyshell itself. As with the last Series I cars, the six-cylinder-engined Series II cars were available with the original (9ft 0.8in/2,764mm) or the lengthened (9ft 4.8in/2,865mm) wheelbase body styles, in

*Five years after the introduction of the original XJ6, Jaguar launched the Series II, which needed only minor retouching. From the front, the main change was the relocation of the front bumper, which made the lower air intakes much more visually prominent than before.*

*From this angle, the Series II car was as graceful, and handsome, as the
original Series I had been. The coloured vinyl roof covering on this particular
car is non-standard.*

both cases with front ends which had been changed in detail.

To meet pending regulations, the front bumper had been raised to a height of 16in (406mm) off the ground, and (on non-USA cars) was accompanied by underriders: Federal models were fitted with massive '5mph' bumpers, the better to meet new low-speed 'no-damage' regulations which related to low-speed accidents.

To match up to this change, the front grille became more squat, while the under-bumper air intakes were more prominent than before. At the rear, the changes were minor, for the original bumper height was retained. Also, with crash protection in mind, sturdy beams were added to the inside of the doors.

Inside the car there had been a complete re-think of ventilation, wiring, instrumentation and general equipment, the result being a fascia which still looked distinctly 'tradi-

tional Jaguar' but which had an entirely new type of air-blending heater, or a different air-conditioning alternative. The instruments and switchgear had been redesigned and relocated, with all dials now placed ahead of the driver's eyes, and with a more logical switchgear layout. In other unseen ways, such as the fitting of a central locking system, the use of multi-plug wiring looms, and reworking all the sound-deadening equipment, the specification was brought completely up to date. There were also space-age innovations such as the adoption of fibre optics for illuminating switchgear.

Mechanically, the big change was a negative one – the dropping of the 2.8-litre engine, which had never been a success. At the time, the company said that it would still be available for certain export markets, but in the event no such cars were produced. The power of the 4.2-litre engine was

*When the Series II cars were launched in 1973, the only visual difference between 'Jaguar' and 'Daimler' types was in the detail of the grilles. HHP 8M is a Daimler Sovereign, while EDU 872M is an XJ6.*

---

## XJ6 Series II (1973–1979)

Specification as for Series I cars (*see* Chapter 2) except for the following:

### Layout
In addition to the four-door saloon, a two-door saloon (badged XJ6C) was built from 1975 to 1977; this car was built on the original length (i.e. shorter) wheelbase underpan

From autumn 1974, all four-door saloons were built on the longer wheelbase

### Basic model range
There was no 2.8-litre engined version of this car

A 3.4-litre model (four-door saloon only) was built from 1975 to 1979

The two-door saloon was sold with the 4.2-litre engine, but never with the 3.4-litre engine

### XJ6 3.4-litre model specification as for 4.2-litre except for:

### Engine
| | |
|---|---|
| Bore and stroke | 83 × 106mm |
| Capacity | 3,442cc |
| Compression ratio | 8.8:1 |
| Max. power (DIN) | 160bhp @ 5,000rpm |
| Max. torque | 189lb/ft @ 3,500rpm |

### Transmission
| | |
|---|---|
| Final drive | 3.54:1 |

---

actually reduced, to 170bhp (DIN), though the customers could not detect this, there was a new engine oil cooler, a new type of exhaust system and other detail changes.

The net result of the upgrade, from Series I to Series II, was to produce a better, safer and rather heavier (by 80lb/36kg) car which in certain respects (driver convenience, and ventilation in particular) was a lot better than before.

## SERIES II DEVELOPMENT CHANGES

During the next five years, Jaguar made a whole series of changes, and genuine improvements, to the Series II cars. Each, in its own way, made an existing car even

better than before. This was the period when a new engine size, a different transmission option, or changes to the trim and decoration could make an impact, and Jaguar played this for all it was worth. There never seemed to be a 'closed season' for Jaguar, as this list of the most important changes made to the six-cylinder-engined cars makes clear.

*February 1974* Overdrive, previously optional, was standardized on the manual-transmission models, with an appropriate pricing change. In practice, almost all recent 4.2-litre cars had been built in this form, so very few customers had cause to complain.

*November 1974* Until this time, the cars had been available with a choice of wheelbases. Although the difference – 4in (102mm) – looks small enough, it seemed to

*From the side view, the Series II car was virtually indistinguishable from the original Series I.*

*Sir William Lyons took infinite care over the detail styling of his cars. This is the rear quarter of the XJ6 Series II, showing careful positioning of chrome strips, decoration, filler caps and glass outlines.*

*An owner's touch – alongside the XJ6 badge of this Series II, the owner has added the key for a security system.*

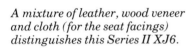

*A mixture of leather, wood veneer and cloth (for the seat facings) distinguishes this Series II XJ6.*

*For the Series II XJ6 models, Jaguar produced a revised and more logically arranged instrument panel.*

*The rear compartment of an early Series II XJ6, this car having mixed cloth and leather for its seat facings.*

make a great difference to the car's appeal. The extra rear seat space made it easier to lounge. Right from the start of Series II production, therefore, sales of longer-wheelbase types far exceeded those of standard-wheelbase types, so from this date the standard-wheelbase saloon car style was dropped. Pressed Steel, for one, was delighted by this, for it reduced the complication of producing Jaguar's bodies.

*June 1975* As an option to the 4.2-litre engine, Jaguar introduced a 3.4-litre derivative, which used the classic 83 ×106mm bore and stroke. History lovers pricked up their ears at this news, supposing it to be a marriage between the latest style and the original type of XK engine – but it wasn't as simple as that. When Jaguar had enlarged the engine to 4.2-litres in 1964 there had been a reshuffle of cylinder bore centres and cylinder block castings; the result was that the 1975-variety of 3.4-litre engine used the latest type of 4.2-litre cylinder block. In the UK, this 161bhp (DIN) machine originally sold for £4,795 (which was £341 cheaper than the 4.2-litre car). Like the larger-engined cars, by this time, it was only available in the longer-wheelbase body.

Although this was what British marketing gurus called an 'entry level' model, the 3.4-litre engined car was still no sluggard – its top speed was 117mph (188kph) – and by some alchemy Jaguar had made the engine smoother and more refined than the 4.2-litre type from which it was developed. It was, incidentally, more fuel-efficient than the 4.2-litre, by two or three mpg. Jaguar expected a lot from this model, especially as it was a 'post-energy crisis' derivative, but must have been disappointed that only 6,880 Series II examples were sold in four years.

*February 1978* For sale only in the USA, where emissions regulations were even tighter than before, the 4.2-litre engine was given Bosch-Lucas fuel injection and a

### Frank Raymond Wilton England (born 1911)

Born Frank Raymond Wilton England, but invariably known as 'Lofty' because of his tall, slim, build, this London-born engineer became Jaguar's Chairman in succession to Sir William Lyons in 1972. Unhappily, he only held this post for two years before the pressures of working in a badly managed British Leyland combine led him to retire to Austria at 63 years of age.

Lofty was educated at Christ College in Finchley before serving a five-year apprenticeship at Daimler's depot in Hendon, North London. He then became a racing mechanic for Sir Henry Birkin, and served Whitney Straight in a similar capacity. After moving to ERA in Bourne in the mid-1930s, he then moved to Coventry in 1938 to join Alvis as a Service Engineer.

During the War he became an RAF bomber pilot, flying many raids in Lancaster bombers, returned briefly to Alvis in 1945, then joined Jaguar as Service Manager in 1946. For the rest of his career – twenty-eight years – he remained at Jaguar, becoming Service Director in 1956, and Assistant Managing Director in 1961. He also combined these posts with the job of Competitions Manager, masterminding the great triumphs achieved by the works C-types and D-types at Le Mans and other circuits all over the world.

In the 1960s, Sir William Lyons singled him out as his successor, but this was a gradual process. He became Deputy Managing Director in 1966, and moved up to become Joint Managing Director in 1967. When Sir William retired in 1972, Lofty became Chairman, but by this time Jaguar was under the thumb of British Leyland. Lord Stokes imposed a 34-year-old Managing Director, Geoffrey Robinson, on Jaguar in 1973, and within a year Lofty handed over the reins to a man who had absolutely no Jaguar tradition, or experience, in his make-up.

catalytic converter, and with the help of a higher compression ratio (8.0:1 instead of

7.4:1) and larger inlet valves, the power output leapt from 162bhp (DIN) to 178bhp (DIN). This was the first boost to the valve diameter since the *original* XK engine had been introduced in 1948. Jaguar, like every other car maker in Europe, had been suffering from the throttling effect of the rules, so this boost in its power was a real tonic for morale.

## Mounting Problems

All in all, it was amazing that Jaguar could give so much attention to its well-established saloons, as it had to face massive political pressures from all sides, it had to oversee the launch of the XJ-S Coupé (*see* Chapter 5), and it had to grapple with the development problems surrounding the new two-door body shell. Without a dedicated (and sometimes downright stubborn) design and development team headed by personalities like Bob Knight and Harry Mundy, the improvements might never have been completed.

With the two-door cars launched in prototype form, and with fuel-injected versions of the V12 engine still under development, Jaguar saw its respected Chairman, Lofty England, opting to take early retirement. With these cars still not yet ready for production, the new Chairman, Geoffrey Robinson, announced plans to double Jaguar's production capacity, shortly before the parent company (British Leyland) plunged towards bankruptcy.

In 1975, British Leyland found itself effectively nationalized by a left-wing government which had no liking for luxury products. A new plan suggested that Jaguar's identity would soon be submerged, for car assembly was to be carried out at the 'Browns Lane Plant, Large/Specialist Vehicle Operations', while engines were to be built at 'Radford Engines and Transmissions Plant' – with the word 'Jaguar' conspicuous by its absence.

Was it any surprise that deliveries reached their 1970s peak of 32,903 in 1976, but fell away rapidly thereafter?

*This was the original XJ 3.4 Saloon of 1975, complete with longer-wheelbase type of body shell, which had been standardized on XJ6 family saloons. Except for badging details, there was no way to distinguish an XJ3.4 from its larger engined relations.*

The Daimler connection was always discreetly advertised. This is a Series II Sovereign (actually a two-door Coupé), with the same proportions and most of the same decoration as the Jaguar derivative.

The two-door Coupés built between 1975 and 1977 were arguably the most elegant of all the body types derived from the XJ6's basic chassis. These cars ran on the original shorter wheelbase, were available with six-cylinder or V12-engine layouts, and had massive doors with frameless windows. This particular car is a 1975 Daimler with the XK engine.

*Mechanically the two-door Coupés were the same as the equivalent four-door saloons, and the floorpan/chassis was the same as shorter-wheelbase types. Only the cabin and the door/window opening arrangements were unique. This XJ6-C was registered in 1975.*

*Take your pick of six-cylinder Coupés – the white car being a Jaguar, the light green car being a Daimler. Both cars, being Coupés, have vinyl roof coverings, and from this view the Series II nose style (which was the same on the four-door cars) is also obvious.*

*The smart two-door Coupé types were revealed in 1973, but at this stage production was not ready to begin. The first cars were actually delivered in the spring of 1975.*

## TWO-DOOR COUPÉS – A SHORT-LIVED INDULGENCE

By almost any *commercial* standards, the XJ two-door Coupés produced between 1975 and 1977 were failures, but this has not stopped them becoming highly attractive to latter-day collectors. In the 1990s, their appeal is their rarity and their subtly different styling. In that case, let's not look too closely at their slow sales records, but more at their character.

The Coupés were conceived when the Series II cars were being planned, and when it was already clear that two versions of the body platform would be available. Assuming the prices were right the sales force liked the idea of selling two rather different cars using the same basic running gear – spacious four-door saloons on the longer version, and sleek,

sporty, two-door models, to be called Coupés, on the shorter version. The Coupés, it was thought, would appeal particularly to North American buyers.

Mechanically, the two-door cars were identical to the Series II four-door saloons. Not only that, but the standard-wheelbase body platforms were the same, the bodies themselves up to the windscreen were the same, as were the two tail ends. The principal – and very obvious – difference was in the side of the body.

Although the silhouettes, and the roof lines, looked to be the same for both cars, the sides were very different, and featured pillarless construction. Instead of providing four passenger doors, Jaguar provided just two doors (4in (102mm) longer than the standard front doors to give better access to the rear seats), allied to rear quarter windows which could be completely wound down out of sight. Like many other such two-door coupés, both European and North American, these new two-door Jaguars looked sensational with door and rear quarter windows retracted. In that condition, and assuming that the passengers did not mind the draught, it was the next best thing to having a fully convertible XJ6.

Was it easier to dream up the *idea* of a two-door version of the car, than actually to *build* it in numbers? At the time, Jaguar certainly seemed to struggle to turn a great idea into reality, and the first deliveries were much delayed.

When the Series II cars were launched in the autumn of 1973, the two-door types were also unveiled, though Jaguar always made it clear that production would not begin until 1974. As it happened, even that promise was premature, as the first two-door cars did not actually go on sale until April 1975. Jaguar's problem was that the design featured doors without a frame or channel at the rear to support the glass, and the design was arranged so that the face of the door glass sealed onto the rear quarter glass. In the event, it took ages for the engineers to

develop a satisfactory seal between the two glasses, for early cars not only tended to leak air at higher speeds, but to set up a lot of wind noise.

When the new two-door cars went on sale in 1975, they were easily identifiable by the use of black vinyl roof coverings. Somehow, too, they looked smaller and more compact than the saloons, an impression borne out by the facts, for a typical Coupé was about 120lb (54kg) lighter than the equivalent saloon. By that time it was the *only* XJ-based car to be built on the original 9ft 0.8in (2,764mm) wheelbase. Everyone who has driven one agrees that somehow, if only by impression, or sensation, these cars feel smaller, and more nimble than the saloons.

Unhappily, however, they fell into the usual trap – the 'less car for more money' syndrome. As with many other cars of this type (Ford-UK's contemporary Granada Coupé was a perfect example), they cost more than the four-door car on which they were based, yet they offered less accommodation, and were not quite as refined as the saloons. Soon after launch, in mid-1975,

Motor's *excellent cut-away drawing of the XJ12C emphasized the way that the engine bay was packed with machinery, and that the cabin was a compact, by no means spacious, four-seater.*

*As the years passed, XJ6 bonnets had to contain more and more gear to combat pollution and meet new regulations. This is a 1975 example of the 4.2-litre engine.*

*Attention to detail by Sir William Lyons was carried to the smallest items, such as the Daimler badging on the nave plate of the road wheels. The year? 1975.*

*Compare this corner view of the original-type XJ6 model, and you see that for the Series II model the bumper/headlight/sidelamp/turn indicator relationship had changed considerably. One reason was to meet the latest USA impact regulations, where bumper heights had to be virtually standardized.*

*Look at the sheer size of the door on the two-door Coupé model, and you can see that the hinges had an arduous job to do. Compared with the four-door too, there are no frames to the drop window glasses.*

*Everyone now knows that there were sealing problems with the doors and quarter windows of the Coupés, but this should not detract from the excellent proportions achieved with this car. As you might expect, there is not a clashing line, plane, or curve at any point.*

*One of several types of instrument panel used in XJ6s over the years – this was a mid-1970s layout, with the additional 'Daimler' touches of that period.*

these were the comparative figures for Jaguar-badged cars:

| | |
|---|---|
| XJ6 four-door | £5,398 |
| XJ6C two-door | £5,777 |
| XJ5.3 (V12) four-door | £6,794 |
| XJ5.3C two-door | £7,281 |

The Coupés, in any case, could not have gone on sale at a more discouraging time. By 1975, British Leyland had been rescued by the government, and in the next two years Jaguar's identity was gradually but inexorably eroded. As far as production engineers were concerned, the two-door cars were unique and awkward cars to build, while the sales force had to work hard to find customers for them. Perhaps the pundits have made too much of the quality problems associated with this shell, for these were fine cars by almost any standards, and road testers made the usual cooing noises when assessing them. British Leyland tried hard to keep the Coupés in the public eye, not only by approving an ambitious motor racing programme (*see* Chapter 7), but by 'placing' them with the high-profile TV series *The New Avengers*.

In the end, however, the special nature of the Coupés brought about their own downfall, for only about 100 such cars were being produced every week. It was clear that proportionally more saloons could be sold if the Coupés were abandoned, so in November 1977 these smart cars were dropped, four years after they had been previewed, and little more than two years after they had gone on sale. For the record, total sales of all two-door Coupés – Jaguar or Daimler, six-cylinder or V12 cylinder – were as follows:

| | |
|---|---|
| Jaguar XJ6 4.2C | 6,487 |
| Jaguar XJ5.3C (V12) | 1,855 |
| Daimler Sovereign 4.2C | 1,677 |
| Daimler Double-Six C | 407 |
| | |
| Total production (1975–77) | 10,426 |

# 5 XJ-S – A Continuing Story

When Jaguar came to consider the future of its sporting cars for the 1970s, it faced unpalatable decisions. These could be summarized simply, and succinctly; could the E-type be replaced, and *should* the E-type be replaced?

The E-type, in fact, had always been a commercial enigma. Did it make money for Jaguar? Was the company satisfied with its reliability record? Was it *really* the right sort of sporting Jaguar for the market? Few other companies would ever have dared to put the E-type on sale at all. Conceived in the 1950s as a successor to the D-type racing sports car, it had been redesigned for the 1960s as a road car. It had always been complicated to build, it had many irritating shortcomings that the starry-eyed enthusiasts tried to ignore, and as the years passed by it became harder and harder for Jaguar to tailor the car to meet the USA safety regulations which multiplied so fast. Even with the magnificent V12 engine installed, the E-type fell a long way behind the standards needed for success in the 1970s. A much better heating and ventilation system, including air-conditioning, was needed – it was no longer enough for Sir William to suggest that the best form of air-conditioning was an open-top roadster with the hood down!

As early as 1968, therefore, Jaguar's top engineers – notably Bill Heynes and Malcolm Sayer – started to plan for a very different car. In the interim, the V12 engine would be slotted into a modified version of the E-type's structure, but for the 1970s a new body structure would be needed. Malcolm Sayer's first thoughts were for a new 2+2-seater sports car based on the running gear of the new XJ6 range. This proposal eventually became formalized as project XJ27, but very little serious design work began until the Series II (V12) E-type was safely launched in 1971.

The big decision, and one of the first to be made at the 'paper project' stage, was that the new car *had* to be a closed coupé – there would be no open-air version. Proposed USA legislation looked like outlawing open-topped cars during the 1970s, on the grounds that roll-over accidents led to many more occupant injuries if a soft-top was fitted. Jaguar, like Triumph, Fiat, Lancia, Nissan and others with the same marketing interests, therefore decided to produce a coupé. Every one of those companies was infuriated to see that the proposed new laws were thrown out in 1974, but by that time it was too late to change plans, and to stop the preparation of production tooling. By 1975, Jaguar, like Triumph with the TR7, was stuck with a design 'frozen' in 1972.

## XJ-S – LAYING THE FOUNDATIONS

By 1971, Sir William Lyons, Lofty England, and technical chief Bob Knight had settled their plans to expand the XJ6 range. It was going to make life rather complicated for production planners in the years to come.

In 1968, the new car had been launched

*The XJ-S ushered in a new type of Jaguar face in 1975, complete with large rectangular-type headlamps and a shallow, almost non-existent grille. This was a much larger car than the E-type for which it was an effective, if not a direct, replacement.*

*The most controversial styling points of the XJ-S, which were never deleted, even at face-lift time in 1991, were the 'flying buttress' panels on the rear quarters. These were originally included for good aerodynamic reasons, but were by no means universally liked.*

with one type of engine (XK 6-cylinder) in one type of structure (a four-door saloon with a 9ft 0.8in (2,764mm) wheelbase). By 1971, Jaguar was determined to use two entirely different types of engine, three wheelbases, three different bodyshells, and two radiator badges. Among all the permutations this would give, there were to be the two-door Coupés, which were really saloons, and another, unique-looking coupé, then coded XJ27, but later to become familiar to the world as the XJ-S.

If Jaguar had still been independent, and

if Pressed Steel had still been an independent body maker, XJ27 might have turned into XJ-S earlier than it did. The fact that Jaguar was now a part of British Leyland, and that there seemed to be an inexhaustible supply of committees and policy bodies to preside over the unwieldy corporation, delayed things considerably. Then, just as the specification was finalized, the world was plunged into the energy crisis of 1973. By that time, in any case, the die was cast. Like all its other new models for the 1970s, Jaguar's principal sporting car was to be a

# XJ-S (1975–1981) (UK specification)

## Layout
Unit-construction steel monocoque four-seater, front engine/rear-drive, with coupé roof and two passenger doors

## Engine

| | |
|---|---|
| Block material | Cast aluminium |
| Head material | Cast aluminium |
| Cylinders | 12 in 60-degree vee |
| Cooling | Water |
| Bore and stroke | 90 × 70mm |
| Capacity | 5,343cc |
| Main bearings | 7 |
| Valves | 2 per cylinder; single OHC operation |
| Compression ratio | 9.0:1 |
| Injection | Bosch-Lucas fuel injection |
| Max. power (DIN) | 285bhp @ 5,500rpm |
| Max. torque | 294lb/ft @ 3,500rpm |

## Transmission
(Manual gearbox only available until 1979)

| | |
|---|---|
| Clutch | Single dry plate; diaphragm spring; hydraulically operated |

## Internal gearbox ratios

| | |
|---|---|
| Top | 1.00:1 |
| 3rd | 1.389:1 |
| 2nd | 1.905:1 |
| 1st | 3.238:1 |
| Reverse | 3.428:1 |
| Final drive | 3.07:1 |

1975–1977: Optional Borg Warner automatic transmission, with torque converter

## Internal ratios

| | |
|---|---|
| Top | 1.00:1 |
| 2nd | 1.45:1 |
| 1st | 2.40:1 |
| Reverse | 2.09:1 |
| Maximum converter multiplication | 2.0:1 |
| Final drive | 3.07:1 |

From April 1977: Optional GM400 automatic transmission, with torque converter

## Internal ratios

| | |
|---|---|
| Top | 1.00:1 |
| 2nd | 1.48:1 |
| 1st | 2.48:1 |
| Reverse | 2.09:1 |
| Maximum converter multiplication | 2.0:1 |

## Suspension and steering

| | |
|---|---|
| Front | Independent by coil springs, wishbones, anti-roll bar and telescopic dampers |
| Rear | Independent, by double coil springs, lower wishbones, fixed length drive shafts, radius arms and twin telescopic dampers |

| Steering | Rack-and-pinion, power-assisted |
|---|---|
| Tyres | E70 205VR-15in |
| Wheels | Cast aluminium disc |
| Rim width | 6.0in |

## Brakes

| Type | Disc brakes at front and rear |
|---|---|
| Size | 11.8in diameter front discs, 10.4in rear discs, with vacuum servo assistance |

## Dimensions (in/mm)

| Track | |
|---|---|
|   Front | 58/1,473 |
|   Rear | 58.3/1,481 |
| Wheelbase | 102/2,591 |
| Overall length | 191.7/4,869 |
| Overall width | 70.6/1,793 |
| Overall height | 50/1,270 |
| Unladen weight | 3,859lb/1,750kg |

### XJS-HE (1981–1991)

Specification as for original model except for:

## Engine

| Compression ratio | 12.5:1 |
|---|---|
| Max. power (DIN) | 299bhp @ 5,500rpm |
| Max. torque | 318lb/ft @ 3,000rpm |

## Transmission

| Final Drive | 2.88:1 |
|---|---|

## Suspension, steering and brakes

| Tyres | 215/70VR15in |
|---|---|
| Rim width | 6.5in |

### XJ-S HE Cabriolet (1985–1988)

Specification as for HE Coupé, except for two-seater Cabriolet style, with removable roof panels

### XJ-S HE Convertible (1988–1991)

Specification as for HE Cabriolet, except for two-seater fully convertible style, and:

## Engine

| Max. power (DIN) | 291bhp @ 5,500rpm |
|---|---|
| Max. torque | 317lb/ft @ 3,000rpm |

## Suspension, steering and brakes

| Tyres | 235/60VR15in |
|---|---|

## Dimensions

| Unladen weight | 4,055lb/1,835kg |
|---|---|

### XJ-S, revised model (introduced 1991)

Specifications as for late-1980s XJ-S HE except for revised style, with new side-window profile on Coupé, and new tail style/tail lamps on all models larger, somehow 'softer', and more versatile machine, based on the platform of the successful new XJ6 model. Marketing analysis led management to decide on a generous 2+2 seating layout inside a closed two-door cabin.

Compared with the E-type, XJ27 was to be a lot larger *and* heavier, with no soft-top option, no hatchback feature, and no wire-spoke wheels. It would have air-conditioning as standard equipment, it would only be available with the V12 engine, and it was expected that the vast majority of cars would be ordered with automatic transmission.

As with all cars that Sir William Lyons was to have a hand in styling, the first decision was to settle on the size of the car's platform which, in this case, only meant choosing its wheelbase. For all the usual product planning reasons, a 9ft 4.8in (2,865mm) wheelbase was chosen for the four-door saloons, and the *original* 9ft 0.8in (2,764mm) wheelbase for the two-door Coupés. Because it was only to be a 2+2 seater, XJ27 could be significantly shorter than the Coupé types, which explains why it was decided to shorten the wheelbase yet again. For XJ27, therefore, a wheelbase of 8ft 6in (2,591mm) was chosen. This, incidentally, was 3in (76mm) *shorter* than that of the Series III E-type, but the new car was so much better packaged than before that it offered more interior accommodation *and* more luggage space. It used a modified version of the existing XJ6 platform, shortened between the seats.

Styling and packaging the new car was tackled in the same way as with other series-production Jaguars of the day. Whereas the E-type had been a refined

*The original XJ-S was a wide, squat, purposeful, very capable indeed, and compact four-seater coupé. Only the leaping Jaguar on the bonnet (placed there by a proud owner) is non-standard on this beautifully preserved car.*

*The XJ-S was a big car, and always looked it, even when it was very carefully posed in a garden setting. This was one of Jaguar's publicity shots of 1975.*

version of a racing car shape (which meant that Malcolm Sayer had by far the biggest influence on its shape), the XJ-S was something of a hybrid. Once again, Malcolm Sayer made aerodynamic calculations and suggested shapes (some of which evolved from those proposed for the mid-engined XJ13 racing sports car of the 1960s), but it was Sir William who then refined these, using his long-established full-size mock-up approach. Sayer, tragically, died in 1970 while project work on this car was still going ahead, and the XJ-S, in fact, was the last styling task to be tackled by Sir William before he retired in 1972.

It was Malcolm Sayer's original sketches and models which saw the birth of the 'flying buttresses' which were to cause such a lot of

discussion in future years, but it was Sir William who retained them, and who also chose to use large ovoid headlamps for all except USA-market cars. Cynics who harp on about those two features ignore the fact that the rest of the style was remarkably graceful for such a bulky car. In any case, when Jaguar stylists later proposed to delete the buttresses, they encountered resistance from the marketing staff, and from the potential customers.

Once the decisions to use a shortened XJ6 platform, the V12 engine, and the relatively roomy 2+2 seating layout had been taken, the general bulk and proportion of the new car was set. A slimmer car could not have used XJ6/XJ12 carry-over components, and would have been less capacious.

|  | **XJ-S** | **E-type Series III** |
|---|---|---|
| Length | 15ft 11.7in (4,869mm) | 15ft 4.5in (4,686mm) |
| Width | 5ft 10.6in (1,793mm) | 5ft 6in (1,676mm) |
| Height | 4ft 2in (1,270mm) | 4ft 3in (1,295mm) |
| Wheelbase | 8ft 6in (2,591mm) | 8ft 9in (2,667mm) |
| Widest track | 4ft 10.6in (1,488mm) | 4ft 6.5in (1,384mm) |
| 'Across shoulders' cabin width | 55.5in (1,410mm) | 48.0in (1,219mm) |
| Front-seat-to-rear-seat dimension (max). | 24.5in (622mm) | 22in (559mm) |
| Unladen weight | 3,900lb (1,769kg) | 3,300lb (1,497kg) |

There was another important difference, however, where the new XJ-S positively shone – that of the aerodynamic drag. To the amazement of almost everyone, when the XJ-S was introduced Jaguar revealed that its drag coefficient was actually less than that of the E-type. At first, of course, this did not seem to be credible. Had not acres of newsprint been expended in the past to praise the sleek shape, and the so-called aerodynamic excellence, of the E-type? It was, the pundits reminded us, a direct descendent of the old D-type, and that had been amazingly slippery too.

Or so the legend had it. In fact, the truth about the D-type, which was less favourable than the legend, had been known for some years before I emphasized it once again in a

---

### Malcolm Sayer (1916–1970)

In the 1950s, Jaguar's racing cars were as famous for their shapes as for their engines. Throughout this period their shapes, and some of their structures, were the work of a large, charming, but ultimately shy engineer, Malcolm Sayer.

Sayer was Norfolk-born, and educated at Great Yarmouth Grammar School, going on to work at Loughborough College. After joining the Bristol Aeroplane Company in 1938, he spent ten years in that illustrious concern, where he gained a valuable knowledge of aerodynamics, and of lightweight structures.

After spending a short time at Baghdad University, helping to set up the engineering facility (and, incidentally, learning Arabic), he returned to the UK, to join Jaguar's small but hard-working engineering team.

Technical chief William Heynes soon recognized his unique blend of talents, the result being that Sayer became the company's one and only aerodynamicist, who used his many contacts in the aircraft industry to ensure that scale models, and even full-size cars, were regularly wind-tunnel tested to allow optimum shapes to be developed.

All the Sayer-designed body shapes – the C-type, D-type, and E-type sports cars being the most famous – were honed using an attractive blend of artistry and mathematics. He knew what had to be accommodated under the skin, made sure that the shape was aesthetically pleasing, yet also ensured that the aerodynamic drag and lift were as low as practically possible.

In the days before computers, all of his calculations were done with logarithms, which involved hours and hours of eye-straining work, and the compilation of huge charts.

The author worked close to him in the design offices of Jaguar for three years, finding him friendly but reticent, helpful but above all secretive of his methods. No one seemed to dare ask how it was done, and Malcolm certainly never offered to enlighten them!

He was a large man, and a heavy smoker, so perhaps it was inevitable that he died of heart disease when he was only 54 years old. His last project, to smooth the lines of the forthcoming XJ-S had only just begun.

*Right from the start, the XJ-S was fitted with a fuel-injected version of the V12 engine, in a much tidier installation than that seen on earlier carburettor-equipped cars. Then, as ever, it was a 5.3-litre unit.*

*The badge, and the large wrap-around tail lamp clusters, were all one needed to know to identify this distinctive Jaguar.*

*The fascia/instrument panel of the original-style XJ-S was well-equipped, but somehow sombre compared with the more glossy XJ saloons of the period.*

*Rectangular headlamps were fitted to XJ-Ss sent all round the world except to North America, where twin circular lamps were fitted instead.*

book in 1983. At the time the new-generation Audi 100 had just appeared with a drag coefficient (Cd) of 0.30 proudly emblazoned on its rear quarter windows, so it was something of a shock to be reminded that the short-nose D-type had had a Cd of 0.500, and the long-nose car, while better, was still 0.489. In 1975 there was almost an *overall* balance between the new and the old. The last of the E-type coupés had a Cd of 0.455, with a frontal area (A) of 17.8sq ft (1.65m²), while the new XJ-S, though looking far craggier, and with a 19.8sq ft (1.84m²) frontal area, had a better drag coefficient of a mere 0.39.

When the XJ-S was first shown to the press in 1975, I can clearly recall Bob Knight spelling out these figures, in his own dry, precise, but totally confident manner. There was general incredulity, but he confirmed them, was clearly proud of the efficiency of the new style, and pointed out that Cd × A for the XJ-S was 7.72, compared with 8.10 for the last of the E-types. The new car, in fact, not only had a better shape, but in spite of its bigger bulk, it was a more efficient package.

## XJ-S ON SALE

Once the style had been settled, development was as rapid as finance and tooling considerations would allow. One car was certainly built by 1971, and (when I was still working in the motor industry) I recall a 1972 visit to the old experimental shops to see a wooden mock-up of the interior, where trim, instrumentation and seat belt fittings were being finalized.

Mechanically, the XJ-S was a very close relative indeed to the XJ12 saloon, with the additional attraction of being available with automatic *or* four-speed manual transmission. Like the XJ12 saloon, it was given a fuel-injected engine, in exactly the same tune, which meant that it produced a magnificent 285bhp (DIN), and massive torque of

294lb ft at 3,500rpm. However, to take account of the better aerodynamic shape, and the possibility of higher top speeds, the final drive ratio was higher – 3.07:1 compared with 3.31:1.

The suspension layout and the power steering were all basically like those of the XJ12, though the XJ-S had a higher steering ratio and the spring and damper settings had been subtly revised. Not only that, but the XJ-S was equipped with a new type of Dunlop tyre which enjoyed the long-winded title of Formula 70 SP Super Sport, this being a 205-section component mounted on newly styled 15in aluminium alloy wheels, with 6in rims.

The engine bay was well filled and, to be honest, offered even less access for servicing than the E-type had done. The major difference between the E-type and the XJ-S, however, was in the cabin, where the accent was now on 'Grand Touring' and space, rather than on a slim cockpit.

I am a tall man, not usually overawed by automotive machinery, but when I first drove the XJ-S, in a quick day's blast around the Cotswolds, to and from Browns Lane, it felt rather large, certainly rather wide, and – at first – not nearly as easy to point and steer as the E-type had been. It was only after an hour's motoring that I began to realize how much 'longer' was the ride of the XJ-S, how stable and secure it felt on undulating roads, and how fast it could be driven on minor roads where visibility was good.

There was no disguising the fact that it was a much bulkier, certainly a much heavier, car than the E-type had been. My first heart-in-mouth experience came when I pressed it very hard around a sharp junction in the hills, applied the massive power a split second too early, and had to deal with a quick and rather unpleasant tail-end twitch. There was a major difference in feel between the E-type and the XJ-S. With the E-type, somehow one rode it, rather like a spirited motorcycle, whereas with the XJ-S one settled down into a very plush leather interior, and let the car do the work. The E-type had to be driven, and to be respected while one was doing it – up to the limits of its tyres, the XJ-S felt much more capable, and much more forgiving.

In the XJ-S, of course, there was a lot more space – more elbow room and more space to settle down and look around – than in the E-type. Right to the end, somehow, an E-type had felt like a sports car that sometimes had a roof, while the XJ-S was definitely a Grand Tourer which had civilized itself to the nth degree.

But would it sell? We had to wait until the end of 1975 to find out. In the meantime, the V12 engine had been fitted to the E-type, to give it a final lease of life; even so, the old car was dying on its feet by 1973. The last of the closed coupés was built in that year, after which open roadster assembly slumped considerably. Monthly production fell from 500+ at the start of 1974, dropped to a mere 107 cars in July, and to twelve in September. The last car of all was produced in September 1974, though the 'official' demise of the car was not announced until February 1975. At the time Jaguar staff freely admitted that a new model was on the way, but they were not giving away any secrets. The new XJ-S was not ready to go on sale, for at that time tooling for the new car's monocoque was not complete, and not even pilot production had started.

XJ-S structures were built on new facilities at the Castle Bromwich body plant; at this time it was still a Pressed Steel Fisher (British Leyland) factory, but in the 1980s it would be taken over completely by Jaguar. There was a new assembly track for the XJ-S at Browns Lane, with initial production set at sixty cars a week, though a rate of 150 cars a week was forecast for later years.

In the event, the new car was launched in September 1975, and first deliveries were made before the end of that year. Most cars had the same nose style, with the large ovoid headlamps, but for the USA twin-headlamp units were specified in their place.

As expected, the new car had a controversial reception. Jaguar's own advertising was headlined 'September 10 1975. A black day for Modena, Stuttgart and Turin', with the sign-off line 'The car everyone dreams of. But very very few can ever own.' *Autocar*'s description was headlined 'A new concept in Jaguar motoring', while *Motor* ran out of new ideas and simply wrote of the 'Return of the Big Cat'.

It was, after all, a luxury coupé rather than a sports car, for it had electric windows, air-conditioning, central locking and four seats – all of which had been absent from the last of the E-types. On the other hand, it went on sale in the UK for £8,900 (the last UK list price for an E-type had been a mere £3,743 . . . ), at a time when the V12-engined XJ5.3 saloon cost only £6,794, and when a Mercedes-Benz 450SLC cost £11,271. Was it going to be too expensive to make its mark?

The XJ-S was enthusiastically received by the world's press, but most of the road tests were published before the new car had settled down into regular production. Sales, which had in any case been quite limited at first, peaked early, at 4,020 cars in 1976, then began to fall away. Not only was the car seen as extravagant at a time when petrol prices were rising fast and the motoring environment in the USA was becoming increasingly unfavourable, but the car was gaining a reputation for unreliability. Only 2,405 XJ-S cars were built in 1979 (and hundreds were still in stock at dealerships all around the world), but a mere 1,057 were produced in 1980. That was the year, incidentally, when XJ-S build was actually stopped for more than six months, and rumours spread that the 2+2 Coupé would soon be dropped altogether. If it had not been for the 'Egan revolution' which followed, those rumours *might* have become fact.

Even so, the design of the car was progressively changed and, in the eyes of the engineers, improved. The original Borg Warner Type 12 automatic transmission gave way to the more modern General Motors GM400 type in 1977, and in 1979 the manual transmission option was dropped altogether. There were cosmetic improvements for the 1978 model year, and at the same time the short-lived XJ Coupés (two-door versions of the saloons) were dropped. Changes to the Bosch-Lucas fuel injection arrived for 1980, while USA-specification cars received a three-way catalyst in the exhaust system; at that juncture there was a much-needed boost to power outputs, from 244bhp to 262bhp for USA-market cars, from 285bhp to 300bhp for 'Rest of the World' cars. Even so, Jaguar's economic traumas affected the XJ-S very severely, and it was only the arrival of a considerably updated model in mid-1981 that helped to save its skin.

## FIREBALL – 'HE' FOR HIGH EFFICIENCY

When the XJ-S HE was introduced in July 1981, John Egan's management was effectively giving the car one last chance to succeed. In the first six years, after all, only 14,972 XJ-Ss had been built, which was nowhere near enough to pay off the substantial investment made in it. Although several important style changes were made to the car at this point, the most important innovation was the adoption of the redesigned V12 engine, complete with the 'May-head' HE (High Efficiency) improvements which I have already described in Chapter 3.

*HE versions of the XJ-S, introduced in 1981, looked much the same as before, except for the new-style alloy wheels, but had a more powerful, more fuel-efficient, engine under the bonnet.*

*The transition from XJ-S to XJ-S HE was mainly mechanical, though at the same time there were slight styling revisions, including the alloy wheels. This car was pictured in original mid-1981 HE condition.*

Those changes, which still feature in Jaguar V12s manufactured in the 1990s, helped the efficiency of the engine, and pushed up the horsepower a little, but they were only one aspect of a package of improvements made to the XJ-S at that time. It was enough to reassure the public that Jaguar had faith in its 'gas-guzzling' coupé, and was intent on promoting it throughout the 1980s.

Like the XJ5.3 saloon, the HE-engined XJ-S had higher overall gearing (which, since one could barely hear the engine in motion, the average owner rarely noticed!) and all cars were equipped with GM400 automatic transmission, while style changes included the use of rubber-faced wrap-around bumpers, wider-rim (6.5in) cast alloy

wheels, and the addition of burr walnut to fascia and door cappings. New instruments, an SIII (saloon) type of steering wheel and updates to items as far removed as the air-conditioning, the switchgear, and the stereo radio/cassette installation all added up to a better package. Perhaps the most important news of all, as far as the customer was concerned, was that the XJ-S HE was £813 *cheaper* than the previous type. It was just what was needed to kick-start the XJ-S's image, for a lot more customers were apparently ready to pay £18,950 for a car which was better and faster than before.

Thus rejuvenated, the XJ-S began to restore its image. Only 1,292 cars (few of them being pre-HE types) were built in 1981, which was awful, but production then leapt

*The XJ-S HE had a squashily comfortable driving position, and in spite of the
150mph-plus performance, the driver was still surrounded by leather, wood
veneer, and full air-conditioning. All V12-engined XJ-Ss, of course, were fitted
with automatic transmission.*

to no less than 3,472 cars in 1982 (with 1,196
sold in the UK), and to 4,749 (1,388 UK
sales) in the following year. That was its
best year, so far. Along with new engine and

new body options soon to be introduced, the
XJ-S was now on a roll, which was to be
sustained until the end of the 1980s.

No changes at the tail of the HE version of the XJ-S, except for the badging. The registration number on this car is very desirable indeed for status-conscious British buyers.

A magnificently maintained XJ-S HE, with yards-deep polish in the paintwork.

*Jaguar was so proud of the efficiency breakthrough made with the 1980s-style V12 engine that it emblazoned the letters 'HE' (High Efficiency) on the boot lid of the XJ-S which took the new unit.*

*Don't be misled by the small grille of the 300bhp XJ-S HE Coupé, for there is also a large and very effective air intake in the gloom under the bumper.*

## CABRIOLET TIME – AND A SIX-CYLINDER ENGINE OPTION

Eight years after the original XJ-S had been launched, Jaguar finally revealed the first body option – not a convertible, as everyone had forecast, but a Cabriolet model. Not only that, but Jaguar chose the launch of this XJ-S to unveil its brand-new six-cylinder engine family, the AJ6 type. Although the new car could, indeed, give open-topped motoring, it was not a car in which the entire top could be folded away, so Jaguar dug out a traditional coach builders' name, and called the new car a Cabriolet – or XJSC for short.

If Porsche had been producing the car, it would probably have been called a 'Targa', for the overall result was the same. Like the Porsche 911 Targa, the XJSC retained sturdy frames around its doors, which were braced above and behind the passengers' heads by a cross-bar. In motorsport terms, the effect was to retain a 'roll cage', making the car more reassuring to those nervous of accidents. Not only could the twin roof panels be removed (and stowed in a bag in the boot, or simply left back in the garage), but behind the roll hoop there was a fold-back soft-top, and when this was furled there was the effect of a convertible without the draughts that usually go with such a layout.

*XJ-SC 3.6-litre model cockpit detail confirms the use of a four-speed ZF automatic transmission, a complex air-conditioning installation, and push-push safety switches.*

**The Tickford Connection**

Tickford, the descendant of an independent coach building concern at Newport Pagnell, was absorbed by Aston Martin in the 1950s. In the late 1970s a new consultancy business, Aston Martin Tickford Ltd, was set up, and began working for the British motor industry in many ways.

One important move was to take over an old factory in Bedworth, near Coventry, a building complex which had once made soft-tops for sports cars and convertibles, which was the right size to undertake batch production of special cars for manufacturers who could not cope with such limited numbers. Among its most famous products has been the speedy conversion of 500 Ford Sierra RS Cosworths into RS500 Cosworths, the building of MG Maestro and Rover 800 Turbos, production of 1990s-style Lotus Elan soft-tops, British Rail coach interiors and the modernization and modification of many Ford RS200s before delivery.

Tickford, in fact, joined forces with Jaguar in the styling, design, development, and manufacture of the 'convertible' aspect of the XJ-S Cabriolet. Tickford developed the system, after which it was also the pivot of a complex manufacturing process.

First of all, from late 1983, normal XJ-S bodyshells, not quite complete, were produced at Jaguar's Castle Bromwich plant, after which they were trucked to the Park Sheet Metal Co. in Coventry for conversion into Cabriolet shells. They were then sent on to Jaguar for painting and mechanical assembly before being trucked yet again to Bedworth, where Tickford completed the Targa panels, the fold-away soft-top, and other details.

At first, Jaguar could not cope with finishing off a Cabriolet under its own roof at Browns Lane, but in due course the complexities of this procedure had to be solved. In the winter of 1984–85 Jaguar took the completion job in house, and Tickford played no further part in building these cars.

Because Jaguar expected this to be only a limited-production version of the car, it chose a rather complex method of building it. Bodyshells were initially assembled at Castle Bromwich, coupés except for the lack of a roof and header panel, after which they were trucked to Park Sheet Metal in Coventry. This specialist 'tin-basher' then modified and completed the body shell (a task which included reinforcing the underframe) as a Cabriolet before it was returned to Castle Bromwich for painting. Then, as with other XJ-S types, it was transported to Browns Lane for all the trim and running gear to be added, before making its final journey to Aston Martin Tickford, in Bedworth, for the 'Targa' panel and the soft-top to be added. This was a complex procedure which could only be tolerated until Browns Lane was ready to fit the AMT bits itself, but that did not happen until the end of 1984. In the meantime, only eleven production Cabriolets were built in 1983 and 178 in 1984, though

this rate rose to nearly 1,700 cars in 1986.

The result was a very different looking XJ-S, not only the first to have a soft-top possibility, but the first *not* to have those flying buttresses between the cabin and the corners. Unlike the XJ-S Coupé, the Cabriolet was only a two-seater, with a luggage shelf behind, which concealed lids hiding further stowage lockers. In later years, in fact, TWR provided a private-enterprise conversion to allow occasional rear seats to be put back in this space!

The launch of the Cabriolet was almost submerged by the release of the new 24-valve twin-overhead-cam AJ6 engine at the same time, for Jaguar made no secret of its intention to use this engine in the new-generation Jaguar XJ40 saloon which was to be introduced later. For that reason, I have given more attention to the new aluminium AJ6 engine in Chapter 8. However, in the 1984-model XJ-S – Cabriolet *and* Coupé types – it was a 228bhp 3.6-litre unit,

*The first XJ-S body derivative – the XJ-SC Cabriolet – arrived in 1983. The roll hoop over the seats was a permanent arrangement, and this car was purely a two-seater.*

*XJ-SCs were called Cabriolets, and had a top that could be opened in two ways – one by removing panels over the passengers' heads, the other way by furling back a soft top, but the roll hoop always stayed in place.*

*With its soft-top erect, and the roof panels in place, the XJ-SC looked like a restyled XJ-S Coupé – and some customers preferred it that way.*

*There were six-cylinder and V12-engined versions of the XJ-SC – this 1987-registered V12 has its soft-top erect.*

*By the late 1980s, the V12 engine was looking tidier than ever before, but it was still a daunting sight for a mechanic who had to work on it.*

*Space, but no seating, behind the front seats of the XJ-SC Cabriolet. Seating conversions were marketed by TWR, but leg room was very limited.*

*The badge says XJ-SC – the 'C' standing for Cabriolet body style.*

*The 'office' of the XJ-SC, 1987 style – there would be yet another fascia restyle in 1991.*

*Detail of the XJ-SC, looking for all the world like a two-door coupé – but the top will come off, and the soft-top can be furled if the weather is suitable.*

## Six-cylinder engine models

**XJ-S 3.6 (1983–1991)**
General specification as for mid-1980s V12-engined XJ-S HE except for:

### Layout
Sold as four-seater Coupé or two-seater Cabriolet (with removable roof panel) at first. Cabriolet option discontinued in September 1987. Full convertible option not made available

### Engine

| | |
|---|---|
| Cylinders | 6 in-line |
| Bore and stroke | 91 × 92mm |
| Capacity | 3,590cc |
| Valves | 4 per cylinder; 2 OHC operation |
| Compression ratio | 9.6:1 |
| Max. power (DIN) | 228bhp @ 5,300rpm |
| Max. torque | 240lb/ft @ 4,000rpm |

### Transmission
Five-speed all-synchromesh Getrag gearbox

### Internal ratios

| | |
|---|---|
| Top | 0.76:1 |
| 4th | 1.00:1 |
| 3rd | 1.39:1 |
| 2nd | 2.06:1 |
| 1st | 3.57:1 |
| Reverse | 3.46:1 |
| Final drive | 3.54:1 |

From early 1987: Optional four-speed ZF automatic transmission, with torque converter

### Internal ratios

| | |
|---|---|
| Top | 0.73:1 |
| 3rd | 1.00:1 |
| 2nd | 1.48:1 |
| 1st | 2.48:1 |
| Reverse | 2.09:1 |
| Maximum torque multiplication | 2.0:1 |
| Final drive | 3.54:1 |

### Suspension, steering and brakes

| | |
|---|---|
| Rim width | 6.0in |

### Dimensions (in/mm)

| | |
|---|---|
| Track | |
| Front | 58.3/1,481 |
| Rear | 58.9/1,495 |
| Overall length | 186.8/4,745 |
| Weight, unladen | 3,575lb/1,614kg |

**XJS-4.0 (Introduced 1991)**

Revised style as for 1991 model V12-engined XJ-S. Specification as for 3.6-litre model except for:

**Engine**

| | |
|---|---|
| Bore and stroke | 91 × 102mm |
| Capacity | 3,980cc |
| Compression ratio | 9.5:1 |
| Max. power (DIN) | 223bhp @ 4,750rpm |
| Max. torque | 277lb/ft @ 3,650rpm |

**Suspension, steering and brakes**

| | |
|---|---|
| Rim width | 6.5in |

**Dimensions**

| | |
|---|---|
| Weight, unladen | 3,814lb/1,730kg |

matched to a Getrag five-speed transmission. Not only was the new engine more powerful than the 4.2-litre XK unit, which was never even considered for use in the XJ-S, but this was the first time that Jaguar had ever chosen to fit an imported manual transmission, the first manual-transmission XJ-S since 1979, and there was no automatic transmission option at first.

That gave Jaguar-watchers plenty to chew on for the next few months, but at least they were mollified when it was learned that the six-cylinder XJ-S could reach more than 140mph (225kph), and could still rush up to 100mph (160kph) in less than 20 seconds. All this, and a potential fuel consumption figure of around 18–20mpg (15.7–14.2l/100km), was reassuring.

Within the next few years, in any case, there was a general reshuffle and expansion of the XJ-S range, for a V12-engined Cabriolet was introduced in mid-1985, and four-speed ZF automatic transmission became optional behind the AJ6 engine early in 1987. In the meantime, more improvements had been made to the V12-engined cars. For 1984, a cruise control, a trip computer, an improved radio/cassette, and headlamp wipe/wash were all standardized, while in 1987 there were further equipment updates including electrically heated front seats.

Then, in 1987, Jaguar not only announced the building of the 100,000th V12 engine, but the 50,000th XJ-S model, and a few

months later the handling of the XJ-S was re-specified with different spring/damper settings, and the option of fatter Pirelli tyres on wider wheel rims. By the end of that year, too, the Cabriolet reached the end of its short life, the last of all being a V12-engined car. That was not to say that Jaguar had admitted defeat with its open car – a rather different version was on the way!

## 1988 – CONVERTIBLE REPLACES CABRIOLET

After less than five years, Jaguar dumped the awkward-to-make Cabriolet in favour of a full Convertible, and at the same time dropped the six-cylinder engine option. The Cabriolet had been a talking point, but had only resulted in limited sales; between 1983 and 1987 a total of 5,007 Cabriolets had been produced, only 1,143 of which had six-cylinder engines. In future, soft-top cars would only be built with V12 engines. As ever, a change that looked simple was actually much more complex than that. The Cabriolet's roll-hoop had provided a lot of shell stiffness which would have to be regained when a full Convertible was developed.

Like the Cabriolet, the new car was, of course, still a two-seater, for the power-operated soft-top took up a great deal of space where a rear-seat back-rest would otherwise have been positioned. There was a storage bin behind the front seats. It took three years for the Convertible structure to be developed and tooled, with a lot of the effort being farmed out the specialist concern of Karmann, in Osnabruck, West Germany. Karmann, which already made smaller cabriolets for VW (the Golf) and Ford (the Escort), was the ideal consultant to do this job, for it not only helped with structural reinforcement, but with the design of press tools and new jig assemblies.

There was extra strength in the floorpan, including stout tubes running fore-and-aft through the sills, more tubes in the A posts, stiffer bulkheads, transmission tunnel changes and extra box-sectioning under the seats. Because of this, and because the buttress had been shorn away, compared with the Coupé there were no fewer than 108 completely new panels and forty-eight modified panels – in other words about one-third of this car was different from the old.

The soft-top was raised and lowered by electro-hydraulic means, and was a conventional framed fabric top, but allied to the use of a rigid glass rear window, which came complete with tinting and a heating element. Opening or closing took a mere twelve seconds. There was no hard-top option.

Like all XJ-S types from this point in history, the new Convertible had Teves ABS brakes as standard, air-conditioning, a cruise control, heated seats, and the long-established fascia style. With the soft-top furled it was a car that could be driven at motorway-legal speeds (in the UK) without buffeting the passengers, while with the hood up, and flat out, it was good for nearly 150mph (240kph).

Unlike the Cabriolet, the Convertible was produced in the same straightforward way as the XJ6 saloon. The shell was assembled in a dedicated part of the Castle Bromwich plant east of Birmingham, painted at Castle Bromwich, then transported to Browns Lane for final assembly.

As a strategy, phasing in the Convertible obviously worked. Jaguar had built 9,052 XJ-S types in 1986, but once the Convertible came on stream this rose to no less than 10,356 cars in 1988, and 11,207 cars in 1989. This was the 1980s peak of XJ-S sales, but by that time the Convertible was well on its way to taking 50 per cent of all XJ-S sales. In the 1980s, indeed, the XJ-S had not only rebuilt its reputation in a remarkable way, but it had also rebuilt its market share. As I have already noted, in 1980 (the last year of 'pre-HE' production) a mere 1,057 XJ-Ss had been produced, but this increased ten-fold by the end of the decade. The older the XJ-S design became, it seemed, the more popular it became. By the end of 1990, and before this model's first major face-lift was announced, the total production figure had reached 88,440 cars, of which 78,420 were XJ-S V12s, 9,979 XJ-S 3.6-litre cars, and 41 were 4.0-litre models.

Naturally, the vast majority of all those cars were sold overseas – in the late 1980s between 65 and 75 per cent of all XJ-Ss built were exported. Even so, the XJ6 range, particularly the long-winded evolution of a new-generation model, had taken up Jaguar's attention for far too long. It was now time for the XJ-S to be revised. The result was unveiled in 1991.

## 1991 – FACE-LIFT TIME

In April 1991, the long-awaited and much-rumoured 'face-lift' XJ-S was revealed. Sixteen years after the original type had been launched, the shell was reshaped at the sides and at the rear, the interior was reworked, and a number of hidden improvements were made to the running gear. Even though the new car was closely based on the old, Jaguar stated that the changes had cost about £50 million, of which £4 million had been invested at Castle Bromwich to provide more accurate body-framing facilities. The

*From time to time, specialists tried to improve on the lines of the XJ-S. This was a design by PBB, from Bristol.*

changeover was all done very smoothly, with no hint of a hiatus in deliveries. The main disruption could have been at the Castle Bromwich body plant, but over-production of the run-out models, and a careful introduction of the new types, meant that the dealers, and the customers, never had to drum their fingers and wait.

In its overall size, its 'package', and its layout, there was no change to the design, but Jaguar claimed that no fewer than 40 per cent of the body panels were new, includ-ing the doors, the sills, the roof, the boot and the rear wings. Visually, the obvious differ-ences were all at the sides and the rear, for

there was a different side window profile behind the doors, and a simple new tail lamp cluster with a horizontal motif to replace the less graceful shape of the previous cars. The new car, and its panels, were the first tangi-ble evidence of Jaguar's investment in the new pressings plant – Venture Pressings at Telford, where Jaguar and GKN had set up a new joint-venture on a 'green-field' site to the west of the Birmingham–Coventry area.

As before, most observers thought (and wrote) that the Convertible was the better-looking of the two types. In 1991, as in 1975, there was criticism of the pressed 'flying buttresses' linking the roof to the tail, and surrounding the boot lid – yet Jaguar's design staff insisted that this was what the customers still expected of the car.

Recently, when I talked to designer Colin Holtum in the design centre at Whitley, he insisted that coupés without buttresses *had* been built, and had been 'clinic'd' by a cross-section of motorists who were in the XJ-S buying sector:

The public told us that they preferred the coupé to have its buttresses, so we retained them on the latest car. Without them – well, it just didn't look like an XJ-S, somehow . . .

Jim Randle confirmed this at the launch of the revised car, saying that a simpler style just 'didn't work'.

In some ways, in fact, the car hadn't been altered to make it look better, but so that it could be better made. On the 1991 model there were fewer, larger, pressings, and of course all the lead-loading which had smothered Jaguars of old had long since disappeared. For that reason, it was easy to miss the fact that the latest car had direct-glazed front and rear glass, new headlamps, doors with frameless window glass and larger diameter (optional) lattice alloy road wheels.

As far as the enthusiast was concerned, the good news was that the V12 engine had once again been re-touched, with a new engine management system, and with a catalyst as standard (peak power was back up to 280bhp), while the six-cylinder engine had been enlarged to 4.0-litres, and even with a catalyst as standard it now developed 223bhp; that of course was precisely the tune in which it powered the XJ6 saloon.

Inside the cabin, there was a completely new instrument panel and switchgear, a new steering wheel and column stalks, electronic memory seat positions, better rear seats, a new trip computer, better air-conditioning, all of which – so Jaguar claimed – provided an entirely up-to-date driving environment. It was from inside the car, however, that I suddenly realized that quite a proportion of the 'new' three-quarter rear glass was actually masked on the inside by trim panels – so why on earth put it there in the first place? The 1970s-style Mercedes-Benz 350SLC/450SLC had been just the same.

This was also the point at which the six-cylinder XJ-S was made available in countries like Canada, Austria, Sweden, Switzerland, Australia, Japan and South Africa. There was no doubt that these, and existing markets, were getting the best XJ-S so far produced, for the six-cylinder engine was matched by excellent transmissions – manual or automatic – while the V12 was still almost miraculously quiet at all times. Neither of the cars was particularly economical (you could expect no more than 20mpg (14.2l/100km) in day-to-day running of a 4.0-litre six-cylinder-engined type) but they were as environmentally clean as any other car on the road.

The good news, as far as Jaguar junkies were concerned, was that the car was still to be sold at very reasonable prices. The new 5.3-litre-engined Coupé, for instance, was priced at £43,500, and the Convertible at £50,600 – which made one look askance at the £70,090 asked for the Mercedes-Benz 500SL. Was this the last change to be made to the XJ-S? In 1991 Jaguar would say no more than forecasting that it would carry on well into the 1990s . . .

# 6 Series III – Re-touching by Pininfarina

## RESCUING A REPUTATION

If the second-generation XJ6 – the XJ40 project – had gone ahead to any normal schedule, there would never have been a Series III version of the original style. XJ40, as you will see, was originally conceived as a 1977 or 1978 model, but its launch was delayed by *eight years*.

More than ten years after the XJ6 had originally been designed, therefore, Jaguar was obliged to give it a major face-lift without spending a fortune, or without making any major technical changes. The result, unveiled in March 1979, was the Series III model. In connection with the considerable, but still subtle, changes which were made to the body style, there was one major innovation – this was the first Jaguar production car to feature the work of an independent styling consultant. From the 1920s to the early 1970s, every Jaguar except the racing and racing-derived cars had been personally styled by Sir William Lyons. But Sir William had retired in 1972, even before work on the much-delayed XJ40 got under way, and before the Series III became necessary.

At the end of the 1960s, with Jaguar as a somewhat reluctant member of the British Leyland colossus, with Sir William about to retire, and with Malcolm Sayer so tragically dead, the company was faced with a problem. Where, and how, were its new styles to be developed? One of British Leyland's

many bright ideas was that a central styling facility should be developed, but as far as

---

**Lord Stokes – the Architect of British Leyland**

From 1961, when Leyland Motors took control of Standard-Triumph, to 1975, when British Leyland was nationalized, Donald Gresham Stokes (who became Lord Stokes in 1969) was the driving force behind Leyland's expansion in the passenger car industry.

Having joined Leyland Motors as an apprentice, Donald Stokes joined the board of Leyland in 1953, and became Standard-Triumph's Sales Director in 1961.

After becoming Standard-Triumph's Chairman in 1964, he was knighted in 1965, masterminded the mergers with Rover (1966–1967), then BMH (1968). At this moment, therefore, the Jaguar Group became a founding member of British Leyland.

On its formation, Sir Donald became British Leyland's Chief Executive, but he also became the corporation's Chairman at the end of 1968. He remained as Chairman until 1975, but after British Leyland was nationalized he stepped down, becoming Honorary President in 1977, then left the corporation completely.

Lord Stokes' most direct influence at Jaguar was to make sure that one of his own protégés, Geoffrey Robinson, took control at Browns Lane after Sir William Lyons and Lofty England both retired.

*The most important change made to produce the Series III model was to provide a different cabin profile, slightly more angular than before, and with more headroom for rear seat passengers.*

Jaguar's top men like Lofty England or Bob Knight were concerned, this was unthinkable. The very idea of having new Jaguars shaped alongside Triumphs, Rovers or – God forbid – Austins, sent shudders down their spines. British Leyland management seemed to be determined to squeeze all the

(Opposite) *Pininfarina's re-touching of the XJ6, into this Series III guise, involved simplifying the nose, to get rid of various air intakes and auxiliary lamps. The Series III's front end was simpler, and more elegant, than ever before.*

character out of the Jaguar marque and at one stage, it is said, Lord Stokes viewed a proposed new Jaguar style, dismissed it almost at a glance, and said 'The only Jaguar thing I want to see on this car next time I come back is the badge on the front'.

Nevertheless, stubborn resistance from Jaguar management proved to be worth it in the end. It is a fact that the *only* new Jaguar shape developed under the Stokes/Barber British Leyland regime, which actually made it into production, was the XJ-S – and this car had been finalized by Sir William Lyons and his staff before he retired in 1972.

Jaguar enthusiasts have always been relieved about this.

Well before he retired, Sir William had laid plans for the future. His solution was to set up a small styling department at Browns Lane, eventually to take over from him. Doug Thorpe led the team at first, which also included Colin Holtum, Oliver Winterbottom and Chris Greville-Smith. A whole department, please note, would have to replace one gifted man! After he retired, however, the staff could not get down to concentrate on new projects at once, for there was a great deal of work still to be done on the XJ-S, on the two-door Series II

cars, and, in theory, on a new-generation saloon car.

By 1974, therefore, the new styling department was already busy with XJ40 proposals, but it was already clear that this car could not be made ready for a number of years; even if the style could be settled, there was a shortage of capital to commit to new tooling. If the department had been larger, and if so much emphasis had not already been put on the XJ40 project, it could have tackled an XJ6 face-lift.

When Chairman Geoffrey Robinson finally got British Leyland board approval for a Series III car, his styling department

---

### Geoffrey Robinson (born 1939)

After Sir William Lyons had retired, and after Lofty England felt obliged to take early retirement, the next Jaguar Chairman was always likely to be faced with 'Mission Impossible'. Geoffrey Robinson not only tackled this challenge with relish, but aimed to double Jaguar's rate of production during his tenure.

Born in Sheffield, Geoffrey Robinson studied German and Russian at Cambridge University, then gained a fellowship to Yale to read economics before returning to England in 1964.

For the next four years, he worked in the Labour Party's research organization, then joined the Industrial Reorganization Corporation at a time when it was heavily involved in encouraging the merger between Leyland and BMH (which included Jaguar). Immediately after the 1970 General Election, the IRC was closed down, whereupon Robinson joined British Leyland, and soon became Financial Controller. After he had run the negotiations to buy Innocenti, he was sent out to Italy by Lord Stokes to manage the business.

Only eighteen months later, Lord Stokes called him back to the UK, and installed him as Jaguar's Managing Director and Chief Executive, immediately under Lofty England. He was then only 34, and had only been in direct contact with the motor industry for three years. The vastly more experienced Lofty England stayed on for only four months under this Stokes-backed regime before electing to retire, whereupon Robinson became Jaguar's Chairman. Although this came about in the depths of the first energy crisis, the enthusiastic Yorkshireman immediately laid plans to double Jaguar's production capability – to 60,000 cars a year – and pushed ahead with the implementation of the complex Series II programme, which included the two-door XJ coupés, the Vanden Plas derivatives of the latest Daimlers, the completion of development of the new XJ-S model.

In retrospect, this period is seen as a time of turmoil at Jaguar, where many middle and senior managers were demonstrably unhappy. They must have been more unhappy in the early months of 1975, after British Leyland had been taken into state ownership, and after the Ryder Report offered a quite impossibly optimistic vision of the corporation's future.

The Ryder Report proposed Jaguar's complete amalgamation with Leyland Cars, leaving no place for an autonomous Jaguar Chairman and Chief Executive, so Geoffrey Robinson resigned, and left the industry of which he had been a member for only five years.

Later he became a Labour MP for the Coventry constituency which includes Jaguar factories. At the time of writing (1991) he is still a back-bencher.

## XJ6 Series III (1979–1986)

Specification as for Series II cars (*see* Chapter 4) except for the following:

### Layout
All cars were four-door saloons, built on the longer wheelbase

### Basic model range
The car was sold with a choice of 3.4-litre or 4.2-litre engines

### Engine
There was no change to the 3.4-litre engine. However, the 4.2-litre engine was modified as follows:

| | |
|---|---|
| Compression ratio | 8.7:1 |
| Fuel supply | Bosch-Lucas fuel injection |
| Max. power (DIN) | 205bhp @ 5,000rpm |
| Max. torque | 236lb/ft @ 2,750rpm |

### Transmission
There was a new type of five-speed manual gearbox as follows:

### Internal gearbox ratios
| | |
|---|---|
| Top | 0.833:1 |
| 4th | 1.000:1 |
| 3rd | 1.396:1 |
| 2nd | 2.087:1 |
| 1st | 3.321:1 |
| Reverse | 3.428:1 |
| Final drive | 3.54:1 |

### Dimensions
| | |
|---|---|
| Unladen weight | 4,033lb/1,830kg |

then convinced him that it was overworked so, strictly as a one-off solution, the job of reshaping the car was awarded to Pininfarina of Turin, Italy. Even the decision to choose Pininfarina was unexpected, for this was a company which, at the time, was more celebrated for producing dramatically shaped sports cars and coupés than saloons. It had been strongly represented at BMC in the 1960s, but in more recent years its offerings to British Leyland had always been rejected; however, it had produced some interesting early proposals for the XJ40 project, and was chosen on that basis.

Considering the tight financial limits, Pininfarina needed all its experience to produce harmonious lines.

## RE-TOUCHING A MASTERPIECE

Pininfarina was told that the body platform/floorpan was to be unchanged, and that the main structure was to be retained, but that it could consider a different cabin, and that it could also propose new front-end and rear-end treatments just so long as major press-tool changes were not required. The restyle, which was offered for approval by Jaguar in 1974, eventually required capital expenditure to the tune of £7 million, a large sum, but by comparison with what was to follow for the XJ40 not a daunting one. In more buoyant times it would no doubt have gone through 'on the nod', but the request

for approval came when British Leyland was approaching its most traumatic financial period, immediately after the energy crisis of 1973–4. Everything was frozen while the corporation was nationalized, the Ryder regime replaced the Stokes management team, and while Leyland Cars was being set up.

No further progress was made at all until 1976, after which the idea of a Series III 'face-lift' once again became a matter of priority, and a launch date of October 1978 was proposed – five years to the month after the Series II car had been revealed. In the event that launch was further delayed, until March 1979, but in the state of crisis which British Leyland found itself at that time, the miracle was that the car was launched at all.

Other than detail changes – to bumpers, to lighting layouts, and to door handles, Pininfarina made no changes below the waistline of the existing car. Above the waistline, the proportions and the pressings were all new. The front screen and pillars were raked more steeply than before, there was a touch more 'tumble home' of the deepened side windows, which made the roof marginally narrower, while the rear of the cabin was slightly further back, with a more vertical rear window than before. The fact that the front quarter window had been deleted made the front quarter seem altogether more streamlined too.

The effect was subtle, but the practical improvements were definite. Although the interior of the car was really no more roomy than before, it *seemed* to be so because of the greater glass area. There was also a small, but welcome, increase in headroom in the back seats. Inside the car, in any case, there was improved seating, a cruise control as standard, and other details to distinguish

*The Series III kept the same basic proportions as earlier SI and SII models, but with a subtly different cabin style, and a face-lifted front end.*

the new car from the old. At the front of the car, the only changes which Pininfarina had been allowed to make were to the grille, and to standardize larger bumpers which hid shock-absorbing pistons on USA models, while at the rear there were matching bumpers and new and larger tail lamp/indicator clusters. Also introduced at the same time were a thermal-glued windscreen, flush-fitting door handles, headlamp wash/wipe equipment and an optional electric sunroof. It was a comprehensive and thoughtful updating package which, as ever, was to be available in Jaguar or Daimler form.

## BETTER ENGINE, NEW GEARBOX

This was not only a visual face-lift, for under the smart new skin there were two important innovations. One was that 'Rest of the World' cars finally received a fuel-injected engine, the other was that there was a different gearbox with five forward speeds. The fuel-injected version of the 4.2-litre XK engine looked the same as that which had been standard on North American market cars a year earlier, but was considerably more powerful. In Europe, the 1978-model carburettored engine had produced 170bhp (DIN) at 4,500rpm, whereas the new fuel-injected type developed no less than 200bhp (DIN) at 5,000rpm.

At this point you might pause to retort that there was nothing very much to be proud of here, as the original 4.2-litre E-type, which had gone on sale almost fifteen years earlier, had produced 265bhp, so where was the progress? The answer, of course, is that in the 1950s and 1960s Jaguar's power claims were no more, and no less, honest of those of their major rivals, and that it had only been decided to re-rate the various engines to the rigorous DIN standards during the 1970s.

On this basis the *original* 4.2-litre XJ6

engine produced 173bhp (DIN) at 4,750rpm. As a guess, too, one would have to rate the very first 3.4-litre XK engine of 1948 at about 135bhp, and this gives a much more accurate guide to the way things had advanced. In fact, the modern combination of a Bosch-Lucas L-Jetronic fuel injection system, and of better breathing in the cylinder head, probably made this the most efficient XK engine yet seen – if not more powerful than the E-type unit, then very close to it, and Jaguar never admitted to anything!

The effect in the car was startling. Here was an old-type engine which had been thoroughly rejuvenated, and while it could not be persuaded to sing round the scale like the latest short-stroke competition, it was a more willing, more torquey *and* more fuel-efficient XK unit than ever before. The 3.4-litre engine was continued, still with its twin SU carburettors, although as before it was not sold in the USA.

The five-speed gearbox, although an innovation for Jaguar, was by no means new – and even today there are Jaguar diehards who are reluctant to admit to its ancestry. It was, in fact, a developed version of a modern British Leyland corporate 'building block'. It was not, in other words, pure Jaguar, for not only had this '77mm gearbox' originally been designed at Triumph, but it had already been used in cars as various as the Rover 3500 executive hatchback, the Triumph TR8 sports, and the Morgan Plus 8. In the years to come it was also to be adapted for use in the four-wheel-drive Range Rovers and Land Rovers. The '77mm' incidentally, is a measure of the distance separating the mainshaft from the layshaft.

Before Bob Knight and his engineers agreed to specify it for the new Jaguars, a great deal of testing and modification had to take place, so that the boxes eventually fitted to all the other British Leyland cars were better, quieter and more refined than they would otherwise have been – with a stronger layshaft and better bearings!

## Bob Knight

When Bob Knight took over as Jaguar's sole technical chief in 1972, his daunting task was to replace Bill Heynes. It was typical of Knight that he never seemed to flinch from this, for he retained the same genial, patient, rather schoolmasterly manner until the day he retired in 1980. He had been Jaguar's Managing Director for the previous two years, under the enlightened 'Edwardes regime'.

The young Knight worked briefly at Standard before joining Jaguar in 1944, and by the 1950s he had already become the company's Chief Development Engineer, operating in what would now be seen as ludicrously small and ill-equipped conditions at Browns Lane. By that time, he had already figured strongly in chassis design for the C-type racing sports car, and in the next few years it was his fanatical attention to detail, especially with regard to ride quality and refinement, which made the Jaguar XJ models so phenomenally quiet and capable.

After Bill Heynes retired, Bob Knight became Director of Vehicle Engineering, working alongside Walter Hassan, who was in charge of all power unit work. Once Walter had successfully seen the magnificent V12 into production he retired, and from 1972 Bob Knight was in total control.

My abiding memory of Bob Knight is of a polite and civilized man, who often seemed to have a half-smile on his face, who enjoyed the explanation and exposition of his job, and who would have been at home in a lecture theatre. When developing a car like the XJ6 or the XJ-S, he seemed to have infinite patience, and would far rather see a new model launch delayed than to have to approve something which, in his eyes, was not yet ready for sale.

This five-speeder, in fact used fourth ratio as its direct ratio – the equivalent of 'top gear' on the old Jaguar four-speeder, which had been rendered obsolete – while fifth speed was a 0.833:1 overdrive. This became the standard gearbox on 3.4-litre and 4.2-litre engined XJ6s (though it was not available behind the V12 engine), and because of its inbuilt 'overdrive' fifth meant that the electrically switched Laycock overdrive unit was no longer fitted. The net result was that the transmission assembly was considerably lighter and cheaper.

## DECISIONS, DECISIONS . . .

But before the Series III was launched, Jaguar had another mountain to climb. It had to claw back its reputation, and restore its individuality. By 1978, the company's autonomy had all but gone, for it had recently become just one-third of a massive business called Jaguar-Rover-Triumph, which itself was merely a minor partner in the British Leyland combine. Bob Knight, newly appointed as Managing Director of Jaguar, had fought long, hard and successfully to preserve some independence of character, but without yet another change of management, and of business philosophy, he might never have seen the company restored to its former level.

By 1977, the nationalized British Leyland colossus seemed to be in financial, organizational and strategic disarray. The workforce seemed to think that it could always strike its way to better wages and less work, the customers were deserting the showrooms in droves, and morale was as low as it had ever been. As *The Economist* commented in March 1977 'The Ryder plan to save and reconstruct British Leyland is now officially a lame duck. The company is next to be a dying one . . .' A week earlier this publication had also proclaimed that 'For the first time since the rescue operation, the National Enterprise Board and the company's management are seriously considering the possibility of eventually shutting part of the troubled car division.'

*Jaguar grille identification – Series III variety.*

*The most noticeable difference between this Series III style and the Series II which it replaced was the more angular sheet metal in the cabin corner area, and a slight increase in glass area and rear seat headroom.*

*Series III noses were simply but beautifully detailed, with sidelamps built into headlamps, and with turn indicators recessed into the front bumper.*

*The Series III fascia/instrument panel shared some components with the early 1980s XJ-S. The telephone was not a standard fitting.*

Things took a turn for the better in November 1977, when the diminutive Sir Michael Edwardes took over as British Leyland's Chairman, his brief being to make sense of the business, by whatever means he saw to be necessary. In fact, this doughty little man had refused to take on the job until he got that assurance.

Sir Michael, who had already spent time serving on the National Enterprise Board, which was the government body controlling British Leyland and other nationalized concerns, already had a variety of changes in mind. One of them was to give pride and a measure of independence back to the marques. As he later wrote in his fascinating book *Back from the Brink*:

Great names . . . were being subordinated to a Leyland uniformity that was stifling enthusiasm and local pride. In fact the Cars operations were split by function and geography; nowhere did the product names appear in the organizational 'family tree'. . . Jaguar assembly took place in the 'Browns Lane Plant, Large/Specialist Vehicle Operations'. . . In short, the worst type of corporate centralism was at work. I found it stifling.

In the short term, Sir Michael set up Jaguar-Rover-Triumph, if only to set up a holding company to separate these marques from Austin-Morris, but for the long term he tried to get each marque to re-emphasize its identity. In Jaguar's case, however, there were major problems; not only was there a growing labour relations problem, but the latest cars were suffering from poor build quality and their reputation was suffering. As Sir Michael wrote:

In the case of Jaguar we failed to solve its many problems at the first go; the product was not reliable, the paint finish was well below par, and productivity was abysmal. Losses were enormous . . . even without the benefit of proper figures it was obvious that Jaguar was losing a lot of money – losses were running at millions of pounds a year. The attitude problem was enormous; the men on the shop floor, and indeed many of the managers still considered Jaguar to be elite, and their own contribution to be unique. Some managers were more concerned with producing new models and teaching new standards of engineering excellence than with managing the business . . .

It was at this point, in 1980, that Sir Michael persuaded John Egan to join Jaguar as the company's new Chief Executive, but this was only at the second time of asking. As Sir Michael later recalled in *Back from the Brink*:

I suggested to Ray Horrocks and Berry Wilson that John Egan, who had left the company [British Leyland] some years before and whom we had failed to recruit the previous year, might be worth approaching again. Between them they very quickly had John Egan in the top job at Jaguar, reporting to Ray, and within days of his arrival things began to happen.

He rejoined the company in April during a strike at Jaguar over a grading issue and immediately became deeply involved in the negotiations. Bridges were built with the workforce from that first day. Furthermore, John Egan believed what other Jaguar executives would not; that Jaguar's mounting losses made Jaguar's demise a certainty unless the turn-round could be accomplished . . .

In the five years since British Leyland had been nationalized, Jaguar production had slumped from 24,525 in 1976 to a miserable 13,978 in 1979. Even though the Series III model was quite clearly a much better car than its predecessor, the demand did not pick up at all, and in 1980 the situation was desperate.

Quite simply, excellent styling, advanced engineering and superb refinement were not

*Two of the most important personalities ever connected with the Jaguar. On the right, Sir William Lyons, the marque's founder, and on the left Sir John Egan, who led the revival of the company in the 1980s. The cars are (right) a 1937 SS Jaguar and (left) a 1982 Series III XJ6.*

enough. The cars were being built very badly, their product quality sometimes being adjudged appalling, and in particular the standard of paint finish and corrosion protection was poor. John Egan's first priorities were to see the product quality improved, and his newly enlarged public affairs staff made sure the world of motoring knew what was going on. A new paint shop complex at the Castle Bromwich body plant near Birmingham (where, at that time, Jaguar still shared facilities with some other British Leyland products) took time to become established, and it was not until about 1982, when Jaguar took total control, that things settled down.

The quality crusade (and, most impor-

tantly, the publicity given to this crusade) took time to pay off, but before long the tide turned. John Egan was always frank about his problems, and within a year of his appointment was noting 210 'significant' faults. 'Our market surveys told us that solving 150 of them would bring us up to the same quality and reliability rating as our German competitors, Mercedes-Benz and BMW.' At one point, in a public presentation, he announced that Jaguar had re-invented the round tyre . . .

Gradually, but repeatedly, the Series III's image began to improve. Jaguar production rose to 22,046 in 1982, and again to 33,437, and in the same period Series III six-cylinder production rose from 11,562 in 1980, to 17,045 in 1982 and to 24,143 in 1984. When you consider that the cars were little changed in that time, the surge was remarkable.

## SERIES III – AN EIGHT-YEAR CAREER

The XK-engined Series III Jaguar XJ saloons were on sale for more than eight years (March 1979 to May 1987) with sales increasing year on year until the very end, when it was finally overtaken by the new-generation XJ40 range. For most of those years, it was an open secret that Jaguar was developing the new XJ40 cars, yet no one seemed to turn away from the long-established XK-engined types. It was a remarkable tribute to the car's style, its refinement, and above all to its character, and Jaguar was delighted to sell no fewer than 151,806 examples.

In general, independent road testers cooed persistently about the Series III. *Autocar* headlined its 1979 test report 'Still distilled excellence' and wrote that it was '... one of the few jewels the British motor industry has left to show the world', and from time to time there was news of the company's return to profitability, and of an increase in sales.

Even so, the last Series III cars of all, built in 1987, were mechanically almost the same as those which had been launched in 1979, for the 3.4-litre and 4.2-litre engines were retained, as was the choice of five-speed manual or three-speed automatic transmission, though the Borg Warner Type 66 took over from the Type 65 in the early 1980s. There always seemed to be some sort of product action to keep editors and enthusiasts interested. When the HE-specification V12 was launched in mid-1981, there was a general reshuffle of six-cylinder Series III specifications and prices, the result being that the 3.4-litre cars became slightly more expensive, but the price of the 4.2-litre car was reduced by a massive £1,550.

In the next few years, in fact, there was a general reshaping of the range by altering names, this also coinciding with the privatization of the company. In the autumn of 1982, the Daimler Sovereign title was dropped from European-market cars, this car being replaced by a new derivative called Jaguar Sovereign instead. A year later, in the autumn of 1983, there was even more mayhem among UK-market product names, for the Daimler-badged cars lost their Sovereign and their Vanden Plas derivatives, while on the other hand a Jaguar Sovereign appeared on the UK market, costing considerably more than the regular XJ6 4.2-litre – and the Vanden Plas badge was retained for top-specification USA-market cars!

## PRIVATIZATION

In 1983 and 1984, however, the biggest news to come from Browns Lane was the long-running story of increasing profits, rumours of privatization, and an upheaval of personnel. At this time government ministers made no secret of their desire to sell off – or 'privatize' – as many as possible of their state-owned businesses. Since Jaguar, as a business, was looking progressively more attractive to investors, it was an obvious target for this treatment.

As John Egan's drive for better quality took hold, and as sales surged ahead, there was a rapid turn-round in Jaguar's finances. In 1981, at the bottom of the trough, the business had lost £36.3 million, but in 1983 this had been converted into profits (on a rising trend) of £55.9 million, and there was more to come. In the spring of 1984, therefore, the government announced its intention to hive off Jaguar from the rest of BL, and to sell off a self-contained business. This had recently become possible by Jaguar taking full control of the Castle Bromwich body plant near Birmingham. Although this ageing complex (which dated from 1938, when Nuffield

Industries started erecting an aircraft 'sha-dow factory' which later produced thousands of Spitfire fighters and hundreds of Lancaster bombers) was really far too large for Jaguar's use alone, it meant that Jaguar's factories – Browns Lane, Radford and Castle Bromwich – were totally independent of any other BL location.

To float the company in a way which would satisfy the City of London, and keep the workforce happy, it was decided that the then Chairman, Ray Horrocks of BL, would stand down, to be replaced by Hamish Orr-Ewing (who was already Chairman of Rank-Xerox), while John Egan would remain as the com-pany's Chief Executive. At the same time, more than 8,000 of Jaguar's workforce were awarded free shares, worth £450, while those with less service, received £105-worth of stock.

The flotation was remarkably successful, and raised well over £250 million, but the new management structure lasted for a mere eight months. At a time when Jaguar announced its 1984 profits, which were an extremely creditable £91.5 million, Hamish Orr-Ewing stepped down from the chair, and was repla-ced by John Egan. The thrusting Egan was now master of all he surveyed at Jaguar, as Chairman *and* Chief Executive, especially as it was known that Sir William Lyons had always approved of his methods.

It was at this time, however, in February 1985, that Jaguar enthusiasts were sad to hear that Sir William Lyons, the company's founder, had died at the age of 83. He had been active, and interested in everything that concerned Jaguar, until the final months. He had always been bitterly disappointed by the treatment meted out to Jaguar by incompe-tent and unfeeling management at British Leyland, and it was almost as if he had been determined to stay alive until he saw his much-loved business sold back into private hands.

Soon after this, in 1986, the new-generation Jaguar saloon, badged XJ6 but usually known by its project code of XJ40 inside *and* outside Jaguar, was launched, but this was still not quite the end of the Series III models. Production of the V12-engined types con-tinued indefinitely, with final assembly now transferred to the Browns Lane 'pilot plant', a building which had previously housed E-type assembly and, in more recent years, the pilot-production of XJ40 saloons.

After more than 151,000 variously badged Series III cars had been built, the last Series III of all, a 4.2-litre-engined example, was produced in May 1987, and went straight into the company's own Jaguar Daimler Heritage Collection of historic vehicles. This meant that XK engine production was drastically reduced, though not entirely stopped. Up at Radford, old machines, in a rather gloomy workshop, continued to build XK engines for fitment to the Daimler Limousine, for use in Alvis military vehicles and for use as replace-ments. It would take more than a simple model change to kill off a phenomenal engine like the six-cylinder XK; after more than forty years of continuous production, it was still in business.

At Browns Lane, however, there had been a complete revolution. During 1986, the Series III XJ6 was progressively run down, and a new shape of Jaguar saloon began to take over. Series III was on the way out – but XJ40 was on the way in!

# 7  XJs in Motorsport

As the 1970s began, Jaguar still showed few signs of returning to serious motorsport, to rebuild the famous reputation it had founded in the 1950s. Times had changed, motorsport had changed, the company itself had changed and, somehow, there seemed to be no incentive.

Although Jaguar had dabbled with the possibility of a return in the 1960s, every scheme considered had fizzled out. The light-weight E-types, though promising, had not been able to beat the Ferrari GTOs, while the mid-engined XJ13 project, so long delayed in its completion, was uncompetitive by the time the prototype was tested. Even though Jaguar's famous 1950s competitions manager, Lofty England, was to become Jaguar's Chairman in the early 1970s, there was a general reluctance to get the company involved in motorsport once again. Like Mercedes-Benz, Jaguar felt that it had developed a long way from its 1950s image. XJ6s and XJ12s, it was thought, were marketed as luxurious executive cars rather than sports saloons.

That situation might never have changed if Jaguar had not introduced the magnificent V12 engine in 1971. Once it was unveiled, enthusiasts and engine tuners soon absorbed the news that it had evolved from a racing project of the 1950s and 1960s. If that was so, they reasoned, surely it could do the job again in motorsport?

## BROADSPEED

In the early 1970s, Jaguar's answers were simple enough – the production V12 was almost entirely different from the original four-cam unit, the XJ saloons were not meant to be racing cars, and the company did not have the resources to develop special competition machines. But one tuner in particular – Ralph Broad, of Broadspeed – was more insistent than the others. Ralph, whose Warwickshire-based team was at Southam, a mere fifteen miles from the Browns Lane factory, had already worked his wizardry on BMC, Ford and BMW cars and engines, with conspicuous success. He took one look at the new Jaguar V12 engine, and stated flatly that he could extract more than 500bhp from it for racing purposes. Ralph, although habitually over-emphatic in his opinions of machinery and personalities, was not a man to be ignored. Jaguar, he said, should return to touring car racing, at the very least to defeat the might of BMW, Ford and Mercedes-Benz in the European Touring Car Championship. Naturally, he thought he should be chosen to set up and run the team to do that job . . .

After Lofty England demonstrated the XJ13 prototype at Silverstone in 1973, Broad asked him to support a racing programme. The answer was no. After Geoffrey Robinson took over from Lofty, Broad made another approach. Once again he was turned down. Then, in 1975, British Leyland acquired new management, and the climate suddenly changed.

In the seven years that British Leyland had been controlled by Lord Stokes, motorsport programmes and almost everything connected with enthusiastic motoring had been firmly suppressed. When the new 'Ryder regime' took over in 1975, however, attitudes changed considerably. British Leyland's new Chief Executive, Alex Park,

## V12 Engine for Racing

Jaguar's famous V12 engine was originally conceived in the 1950s as a four-overhead-cam racing unit. The first engine, completed in 1964, was a 5-litre which produced 502bhp, but it was not further developed. The single cam V12 which went into production was a very different design in many ways.

The first racing application of the single-cam V12 was developed by Broadspeed in 1976 and 1977; this was a 'Group 2' unit for the XJ-S, and produced about 500bhp. This racing programme was short-lived, but in the USA Bob Tullius' Group 44 team produced successful 476bhp versions to win the Trans-Am Championship.

Tom Walkinshaw then started again with the Group A XJ-S cars in 1982, and by 1984 these engines were producing about 450bhp. Then, as later, many of the parts were manufactured by Cosworth engineering, though most of the actual development was done by TWR itself at Kidlington.

Then came the two interrelated Group C and IMSA racing sports car programmes, in which TWR (Group C) and Bob Tullius' group 44 (IMSA) teams were involved. The TWR achievements give an idea of what was achieved during the 1980s.

The 1985 XJR-5's engine was a 6.2-litre/650bhp unit, the 1986 XJR-6 used a 6.5-litre/700bhp unit, while for 1987 the XJR-8 was a 7.0-litre engine with a similar power output. All these engines used modified versions of the original 'flat-head' single-cam cylinder heads. By this stage, too, TWR had also taken over the IMSA programme in the USA, where engines were limited to 6.0 litres.

By the late 1980s, TWR had also developed a 4-valves per cylinder twin-cam cylinder head for the racing engine, and found that there was still scope for the engine to be further enlarged. For the 1991 Le Mans 24-Hour race, however, the well-proven XJR-12 still used the single-cam head, but the engine had been enlarged to 7.4 litres and developed 730bhp at 7,000rpm. The torque of that engine was an astonishing 610lb ft at 5,500rpm.

The measure of TWR's abilities was that in seven years the basic production V12 had been transformed, had been enlarged from 5.3 litres to 7.4 litres, and had seen its 300bhp road car power figure increased by nearly 250 per cent.

appointed Derek Whittaker to run the Leyland Cars division (which included Jaguar), and before long it was clear that motorsport was back on the agenda.

In March 1976, Leyland Cars called a press conference to announce, with considerable bravado, that Jaguar was to contest the European Touring Car Championship, and that Broadspeed was to run the new Group 2 cars. The impression given was that outright victories were expected almost immediately! At the time, Derek Whittaker made his point briefly, and succinctly:

We are going racing with Jaguar for exactly the same reasons as we do anything else in our business – to sell more cars and make money . . .

Perhaps it was as well that Sir William Lyons had retired from business, for this showbiz type of presentation was contrary to everything in which he believed. Even the cars chosen to do the job – V12-engined XJ5.3Cs (the two-door version of the XJ12 saloon) – carried Leyland Cars logos which were as large as the leaping Jaguar symbols, and were kitted out in a garish red, white and blue colour scheme. Somehow or other a four-speed manual gearbox was homologated for this car (no XJ5.3C *production* car ever had a stick shift . . .). Vast flared wheelarches were fitted at front and rear, to accommodate the 13in rim-width wheels and Dunlop racing tyres.

No one ever satisfactorily explained why the XJ5.3C was chosen for this job, when the

*The Broadspeed racing Jaguar XJ5.3C ready to spring away from the start line of the Tourist Trophy race at Silverstone in October 1976. The car was phenomenally fast, but failed to finish.*

XJ-S Coupé would have been so much more suitable, and could also have been homologated into Group 2. The XJ-S, after all, used the same running gear, but in a smaller and more aerodynamically efficient bodyshell which was about 200lb (90kg) (the weight of the driver, or of 25 gallons of fuel) lighter. The only cautious reason given was that Jaguar (and Broadspeed) was not sure that the XJ-S's cabin was large enough for it to qualify as a four-seater 'touring car'. But if Ford's Capris, and BMW's coupés qualified, surely the Jaguar would have complied?

Work on the new project began in 1975, and when the programme was announced one car, re-engineered in a complex manner, was already being tested. Leyland announced the signing of four star drivers – Derek Bell, David Hobbs, Andy Rouse and Steve Thompson – claimed 480bhp from the Broadspeed-developed engine, and stated that the car would make its début at the Salzburgring only a month into the future.

## Disappointing Results

Unhappily for all concerned, this was the first of many promises and forecasts which

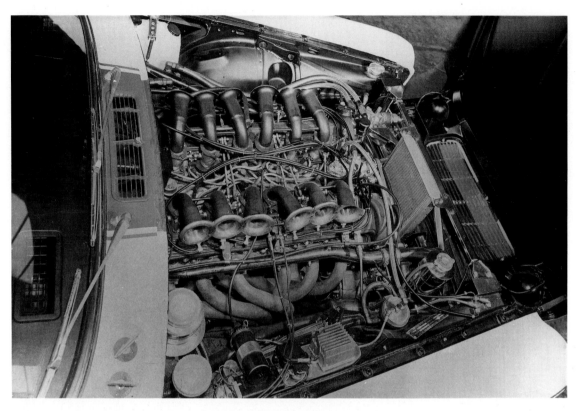

*When the Broadspeed-prepared racing Jaguar XJ5.3C was revealed in March 1976, its V12 engine looked messy – but produced about 500bhp.*

were not met. The project was a sad let-down in 1976, and little more successful in 1977. The Salzburgring début was abandoned when all manner of teething problems showed up in testing, and eventually the car did not race until the British Tourist Trophy race of September 1976, six months after the launch. There was never any doubt about the vast potential of the car – *if* the engine could be persuaded to stay in one piece, and *if* the brakes could cope with an all-up weight (driver and fuel on board) of around 4,000lb (1,814kg).

Because Group 2 regulations of the period did not allow the use of alternative dry-sump lubrication systems for the engines, Broadspeed struggled for months to tame the sump oil surge which was inevitable

under hard braking and acceleration. Water-cooled brake callipers were also outlawed by the regulations, and there were problems with wheels and suspension units breaking up under the strain of test driving, so it was no wonder that morale progressively dropped.

By September, Steve Thompson had left the team, and Chris Craft had been signed up in his place, but there was still only one raceworthy car when the Tourist Trophy came around, at the Silverstone circuit. Derek Bell was fastest in practice, putting the car on pole, and duly led for the first nine laps of the race before a pit stop was needed for a tyre change. Later in the race a wheel came off as a drive shaft disintegrated, and the Broadspeed car's first race ended in

*The racing XJ5.3C started as 'No. 1' and was certainly the fastest car on the track at the 1976 Tourist Trophy race at Silverstone, but it was forced to retire.*

*For 1977, the 'works' XJ5.3Cs, as prepared by Broadspeed, not only sported a different livery, but had a different front spoiler (complete with air intakes for the brakes, and for oil coolers), and had a plastic spoiler across the boot lid.*

*In profile, the Broadspeed racing XJ5.3C looked graceful and purposeful. This was the 1977 specification of the cars, complete with corporate British Leyland livery scheme, and with large rear spoiler.*

retirement. That was its only racing appearance of the year.

Nevertheless for 1977, and after a winter's development, great things were expected of the car. The new driving line-up was Derek Bell, Andy Rouse, John Fitzpatrick and Tim Schenken, and a full-scale assault on the ETC was promised. By this time there was a much modified colour scheme, with more 'Leyland' and less 'Jaguar' emphasis, and a rear spoiler had been homologated – the rules being such that 500 identical spoilers had to be supplied (as 'fit-these-yourself' kits delivered in the boot of new cars) to XJ5.3C customers before these could be approved.

The latest cars had more power than ever before – up to 550bhp – and some weight had been pared away, but the whole unhappy season was to be a battle against unreliability connected with engines and the sheer bulk of the cars. At Monza, oil surge led to engine failures, only one car started the race, and another engine failure followed

after the 'Big Cat' had led the race for an hour. Other makes of car also suffered engine failures due to the same problem. Not long after this, as a consequence, the authorities decided to change the rules to allow dry sumps, but Broadspeed did not finalize its new system until midsummer.

At the Salzburgring, Andy Rouse's car led for a time, but both cars retired with drive shaft failures, they did not start at Mugello because new shafts were not yet available, while at Brno the cars led in the early stages before one burst an oil pipe and the other burst a tyre at high speed. Two cars started the fearsome Nurburgring four-hour race, but one lost its engine almost at once, while the other (driven by Derek Bell and Andy Rouse) finally finished, strongly, in second

(Overleaf) *A Broadspeed XJ5.3C, at rest, showing off its final (1977 season) colour scheme.*

place, but 2.5 minutes behind the winning BMW. At Zandvoort, in August, one car had a dry-sump installation, whose pump drives gave endless problems – and both cars had considerable mechanical problems.

By that time, British Leyland had decided to cancel this programme at the end of the year, and the fate of the cars was already sealed, so what happened at the Silverstone TT in September would be academic. A Jaguar took pole position in practice, both cars led from the start of the race, but Tim Schenken crashed the 'dry sump' car when a front hub broke up. After all the pit stops had been completed the Bell-Rouse car found itself second, 20 seconds behind the leading BMW (one of whose drivers was a young man named Tom Walkinshaw . . .). Rouse is convinced that he would have regained the lead if the weather had not suddenly turned nasty. A combination of a shower, and a patch of someone else's engine oil caused him to spin off and crash the car.

There was one final twitch in the dying programme, with both cars taken to Zolder, in Belgium, where one car blew its engine and the other suffered a seized gearbox. To sum up the two-season saga, the 'works' XJ5.3Cs had started eight races and led them all for a time, but in fifteen starts they had finished only three times.

## TRANS AM RACING

In the meantime, over in the USA Bob Tullius' Group 44 racing team had got to grips much more successfully with XJ-S cars developed for Trans Am Category One racing. In 1977, the team won five times in ten races, with Tullius himself winning the Drivers' Championship, while in 1978 Tullius won again, the team won the Manufacturers' series, and there were seven outright race victories. Then, after a couple of years of inactivity (while Group 44 raced Triumph TR8s instead), Tullius' team produced a

'silhouette' XJ-S to use in the top Trans Am category. In 1981, Tullius won three races, and finished runner-up in the Drivers' series.

The Jaguar factory, however, had been so badly bruised by its failures in 1976 and 1977 that it kept its head down, and refused to get involved in motor racing again for some years. The company, in any case, was in financial disarray at the time. John Egan arrived in 1980 to attempt 'Mission Impossible' with the run-down business, and there was neither time nor money to back a motor racing team.

## TWR

In 1982, however, a new set of motor racing formulae came into force, one of them – Group A – applying to touring car racing and rallying. Compared with the old Group 2 rules, this meant that cars at least had to look exactly like those which the public could buy (which put an end to the use of optional wheelarch flares, spoilers and other aerodynamic aids) and engine tuning was more severely restricted.

Tom Walkinshaw, whose TWR business was based in Kidlington, near Oxford, looked long and hard at the new rules, surveyed Europe's eligible cars, and decided to prepare a Jaguar XJ-S for this new sport. He thought he could make them as fast *and* a lot lighter than the old XJ5.3Cs. This time around, however, he did not depend on Jaguar for his backing, except to make sure that it was homologated with appropriate extras.

Having talked to Sir John Egan, he was first told to go away and start winning before any support could be guaranteed. By mid-1982 TWR had prepared a single black car, whose little-modified engine produced 420bhp, which was backed by Motul and Akai. Jaguar had been persuaded to homologate a five-speed Getrag manual gearbox as an option, which made the car much more competitive than it would have been with automatic transmission. Dry weight was about 3,087lb (1,400kg).

Tom's first major success came in the Brno Grand Prix (Czechoslovakia) of 1982, the second (with Chuck Nicholson as his driving partner) followed in the Nurburgring Six-Hour race, then two further victories follow-ed in the Silverstone and Zolder rounds of the European Touring Car series.

Remarkable! And what a contrast to the humiliatingly public failures of 1976 and 1977. It was no surprise to see Jaguar giving its official backing to the endurance prog-ramme in 1983, where a typical race was 500km long, lasted for about three hours, and required two drivers for each car. In the next four seasons the factory-backed XJ-Ss, with engines using a series of Cosworth-produced components, became *the* class of the field, all over the world. It was a serious, and comprehensive, programme which, in the end, saw a total of seven cars built, and saw the 5.3-litre engine's power output increased to about 480bhp. By the end, the cars were impressively reliable.

In 1983, the cars were white with green 'Motul' striping, and a two-car assault on the ETC saw five victories achieved. Walkin-shaw himself drove three of those winning cars and finished second in the Drivers' Championship. A year later the TWR team swept the board. Running the same cars, but with a new colour scheme (Jaguar racing green with white stripes, but now with no obvious backing from Motul) three cars started most of the races. Not only did TWR win the European series for Jaguar, but Wal-kinshaw won the Drivers' Championship. There were six outright victories during the year, including a crushing success in the gruelling Spa 24-Hour event, other 'winning' drivers including Hans Heyer, Nicholson, Winston Percy and Martin Brundle.

That was enough for Jaguar to decide to withdraw from the ETC as undisputed win-ners, and to back TWR in an even more ambitious programme – that of developing a series of of V12-engined Group C race cars to attack the World Sports Car Championship. The good news for XJ-S enthusiasts, however, was that money was still available to back the XJ-Ss in selected events around the world.

In the next two years, TWR entered the

*Would Sir William Lyons have approved of a Jaguar that looked as extrovert as this? The Broadspeed cars of 1976 and 1977, however, were fast, functional, but ultimately successful variations on the XJ theme.*

*TWR's Group A racing XJ-S Coupés were always competitive, and soon started winning Touring Car races all round the world. This was their final, and famous, 1984 colour scheme.*

*British racing green, with a large white stripe, and with 'Jaguar' proudly emblazoned on the doors, indicates one of the fast, famous, race-winning Group A XJ-S coupés of the early 1980s, when the cars dominated European Touring Car Championship races.*

familiar XJ-S cars in a variety of high-profile events, though its European Touring Car championship efforts were now being placed behind a team of V8-engined Rover Vitesse saloons instead. Two XJ-S cars were sent to Macau in November 1984, where they finished first and second, and less than a year later no fewer than three cars were sent to compete in the world-famous Hardie 1000 saloon car race at Bathurst in Australia. One car retired, but the other two finished first and third overall – an excellent result for a team on its first visit to the celebrated high-speed circuit. It was the Group A XJ-S's twentieth, and final, overall victory.

The XJ-S's homologation ran out at the end of 1986, but after FISA had granted special dispensation for the cars, they were entered in two New Zealand races at the beginning of 1987. Their last outing of all, at Pukekohe, resulted in second place, with Winston Percy and Armin Hahne at the wheel.

Thus it was that TWR's Jaguar XJ-S effort came to a close, almost five years after it had begun. For lovers of tradition, and for those who wanted to feast their eyes on the cars in the future, all seven cars survived into the 1990s. Tom Walkinshaw, who was as nostalgic as the next man, kept some of them, the others being sold off to Jaguar collectors.

# 8 XJ40 – The New Generation

When the second-generation XJ6 saloon was launched in 1986, it ended one of the longest gestation periods in automotive history. The new XJ40 concept – as it was always coded by factory planners – had been under development for no less than thirteen years. This may seem incredible to enthusiasts, but the fact is that the first clay model was produced in 1973. There are grounds for suggesting that this project should be entered in the *Guinness Book of Records*, and I am not at all sure that Jaguar would be proud of that – I certainly cannot recall any other car which took so long to make it to the showrooms.

In the case of the original XJ6 project in the 1960s, the first cars were sold rather less than four years after Sir William Lyons had finalized a style. So, for the new car, what took so long? The lengthy delays, quite simply, can be summed up in two words: British Leyland. The dead hand of corporate inefficiency, petty jealousies, and sheer lack of vision harmed individual marques considerably during the period in which the new Jaguar saloon car was conceived. Lord Stokes and John Barber, who effectively ran British Leyland before it was nationalized, were determined that 'Leyland', not 'Jaguar', 'Rover', 'Triumph' or any other marque, should emerge as the corporation's dominant badge. Things did not really improve after nationalization, for although their management team was replaced by Lord Ryder, Alex Park and Derek Whittaker, the policies didn't change.

Firstly Sir William, then Lofty England, and then Geoffrey Robinson – the three Jaguar chairmen who served in the first half of the 1970s – all realized that the first-generation XJ6 could not live for ever. In spite of a distinct lack of support from the top, therefore, they tried to get a new project going in the first half of the 1970s. The first sketches were made in 1972 – the year in which the XJ12 was launched – but the very first new style, in quarter-scale model form, was completed in October 1973. This looked remarkably like that which was finally approved, except that it had a distinct touch of XJ-S Coupé around the nose. At that stage, however, there was still a lot of work to be done, all manner of individuals and product committees had still to be satisfied, and all that could be agreed was the approximate size of the new car's platform, and that it would eventually have new-generation engines.

In the next few years there were to be many different attempts to settle on a future style, and – for the first time in Jaguar's history – a whole series of consultant offerings by independent design houses were also assessed. It all started in 1973 when Pininfarina bought an XJ12, modified it, showed it at various motor shows around Europe, then delivered it to Browns Lane for viewing. As you might expect from this fashion house, there were distinct Ferrari tendencies in its shape.

In 1973 Geoffrey Robinson who, having worked at Innocenti as Managing Director of that Italian-based concern, knew most of the Italian styling houses, commissioned full-size clay models from Giorgetto Giugiaro (of Ital Design), and from Bertone. (Giugiaro, in fact, produced *two* different full-size clays, one of

*The new generation XJ6 of 1986, in 24-valve 3.6-litre form, shown in cutaway.*

which was eventually refined to become a Maserati saloon car!) Clay after clay was then produced by Jaguar's own studio, in rapid succession, with the shapes gradually tending to drift away from sinuous lines, and towards a harder-edged profile. Ital Design and Bertone tried again in 1976, but after that Jaguar stuck to its own themes, and the rounded lines gradually came back again.

In the meantime, Pininfarina had produced the Series III face-lift on the existing XJ6 platform and body structure and, since this was obviously a great visual success, the pressure to finalize a new car had somehow been removed.

In recent years, Jaguar has released a mass of photographs showing how the style developed, and what strikes me strongly is that the 'first thoughts' shape of 1973–4 was an excellent one which kept recurring over the years. Stylists and artists could no doubt give me many reasons why the *chosen* style which finally went on public view in 1986, was better than that of 1973. I will only reply that

the 1973–4 style was a very good base to refine at the time!

By 1978, the finalized shape was beginning to emerge – yet at this juncture I ought to remind readers that this was still a year before the Series III model made its bow, and eight years before the new-generation car actually went on sale. The encouraging development, too, was that Sir William Lyons, who had kept well away from Jaguar while the corporatist policies of British Leyland were at their height, began to make frequent and very welcome visits to Browns Lane to view styling models. His opinions were carefully noted for, even in his seventies, his eye had not dimmed, and his taste was still unerring.

(Overleaf) *More than ten years of concept, test and proving work on the XJ40 project eventually produced this smooth, some say understated, style. Only the 'entry-level' cars were fitted with twinned headlamps for the British market.*

By 1980, Jaguar had been struggling along under the guidance of Michael Edwardes' management team for more than two years, and found that the XJ40 style had finally reached a point where it could be approved. In spite of all the problems surrounding the company, Bob Knight, his top management and stylists were all satisfied that they had finally developed a worthy successor to the Series III model. In almost his last major executive act – for he retired in mid-1980 – Knight sent the full-size clay model to Pressed Steel Fisher for detail measurement, and for the first drawings to be prepared. If there was ever a make-or-break point for XJ40, this was it, though more than six years were to elapse before the car would go on sale.

## GETTING THE SHOW ON THE ROAD

Jaguar's, and British Leyland's, problems at this time have already been related. At this point I need only recall that the company's annual sales had slumped from 32,478 in 1974 to a miserable 13,978 in 1979. This explains, quite simply, why Jaguar had very little money to spend on new products at the time and why Jim Randle, as Director of Vehicle Engineering, could allocate so few resources to the new XJ40 project. Yet in the spring of 1980 there was a real upheaval at Jaguar, with all manner of industrial and managerial changes. As already described, the workforce went on strike over a grading issue, one so bitter that it could have seen the company closed down for good. John

---

**Sir John Egan (born 1939)**

When John Egan arrived at Jaguar in 1980, the company was virtually on its knees. The assembly lines were at a standstill, the marque's quality reputation was at an all-time low, and it was not expected to survive for long.

In 1980, the company lost £46 million and built a mere 14,000 cars. Six years later, production had rocketed to 43,000 cars, and profits were running at £122 million a year. Not only that, but the company had successfully split from BL, and been privatized; the Stock Market loved every minute of it. Much, if not all, of that improvement was due to the dynamism and leadership given to the company by Egan, who was knighted for his efforts in the Birthday Honours list of 1986.

Before joining Jaguar, Sir John had completed impressive spells with BL's Unipart division, and with Massey Ferguson. At Jaguar he had an uphill struggle, not least in convincing unions and suppliers that he *would* close down the business if quality and labour relations did not improve; but he turned the business round. Helped along by inventive and vigorous public relations activity, he also convinced the world (particularly North America) that Jaguar was on the way back

At first he was Jaguar's Chief Executive, reporting to Ray Horrocks of BL Cars, and for a short time after privatization in 1984 he kept that position under Hamish Orr-Ewing, but after Orr-Ewing stood down in March 1985, Sir John became the undisputed boss of Jaguar. Not only did he bring the new-generation XJ6 model (the XJ40 project) safely to the market in 1986, but he also master-minded the purchase of Whitley, the setting-up of the pressings venture at Telford, and – most important of all for the sporting enthusiasts – he supported Tom Walkinshaw's TWR enterprise on its way to motor racing success.

Until 1988, in fact, he kept every possible ball in the air, but once the long-drawn-out negotiations with General Motors stalled, and the forced sale to Ford had been agreed in 1989, it was clear that he would soon be stepping down. His resignation became effective in June 1990 (when Ford-nominee Bill Hayden took his place), after which he moved on to take control of BAA, the British Airports Authority.

Egan arrived to become the company's Chief Executive, and it was his personality more than any other which caused the strike to be called off shortly afterwards.

After John Egan had settled in, Bob Knight retired. Since Knight's engine design chief Harry Mundy had already retired, this meant that two major new engineering appointments were needed in a very short time. Trevor Crisp, a student at Jaguar in the 1950s when the author also joined the firm, slipped into Mundy's role, while Jim Randle became Engineering Director. Randle, who already had fifteen years' service at Jaguar, had been Jaguar's Director of Vehicle Engineering since 1978, and could not have been appointed at a more dispiriting time. He led a small team which had started project work on XJ40 in 1979, but had been pessimistic about the company's survival in April 1980, and thought the project – his job, even – would be killed off.

Accordingly, he was delighted to be authorized to start an XJ40 build, test and development programme later in that year. His original target was to have the first SEP (semi-engineered prototype) on the road early in 1981, but the first car actually ran in mid-July. The very first car to be tested had the finalized body style, but there were a great many differences between the prototypes built in 1981 and 1982, and the cars which finally went on sale at the end of 1986. In the five years which intervened, Jaguar put in an unprecedented amount of test and development work, not only on test beds and rigs, but on the roads and tracks of the world.

In Europe, the XJ40's natural competition was going to come from BMW's 7-Series and the Mercedes-Benz S-Class range, but Jaguar also knew that in the USA it would be judged against several domestic products, and that it was highly likely that the Japanese were also planning to enter this market sector. The Series III was already setting

## Jim Randle (born 1938)

To succeed Bob Knight as the company's Director of Product Engineering, Jaguar chose another quietly spoken and contemplative engineer, Jim Randle. At the time, in 1980, Randle already had twenty-six years' motor industry experience, the last fifteen of them at Browns Lane.

The Birmingham-born Jim Randle was educated at Waverley Grammar School, then elected not to go on to university, but immediately joined Rover, at Solihull, as an apprentice in 1954.

In the next three decades he not only collected an impressive list of qualifications, many of them achieved in and around his normal work, but he also rose steadily to become one of the industry's most respected engineers. By 1965 he was a Rover Project Engineer. He then joined Jaguar Cars, and went on to become Chief Engineer, Vehicle Research, in the 1970s. There was then a two-year spell as Director, Vehicle Engineering, under Bob Knight, before the final promotion to direct the fortunes of the entire product engineering team.

During the 1980s, Randle not only headed up the teams which designed and developed the new-generation XJ6 (XJ40) models, but his 'Saturday Club' also produced the sensational XJ220 'Supercar' prototype. Many Jaguar enthusiasts still look on him as the 'Keeper of the Flame' of Jaguar tradition, an honorary title which he holds with pride.

After the Ford take-over of 1990, Clive Ennos took over his mainstream product engineering functions, so he could then concentrate on his major enthusiasm, which was the conception and creation of entirely new products. At that time Jim Randle was only 52 years old, which left him many productive years to shape the Jaguars of the future, but with the 'F-Type' cancelled and the new model plan re-drawn yet again, he clearly chafed under the more stringent controls imposed by Ford. In 1991 he left the company, stating that he did not intend to work again in the motor industry.

'Jaguar' and 'XJ6' appear on the tail of this 1991 model, but there is no mention of the engine size – which could be 3.2 litres or 4.0 litres.

Could any design have been carefully integrated, in detail, than the late-1980s XJ6?

Because many Jaguars are bought by
companies, who expect their bosses to ride in
the rear seats, Jaguar had to provide
comfortable seating with ample rear leg
room.

Plenty of comfort, but plenty of function, too,
in this 1991 version of the 4.0-litre XJ6.

By the early 1990s, the new-generation XJ6
had already received a revised fascia, with
the earlier digital read-outs discarded.

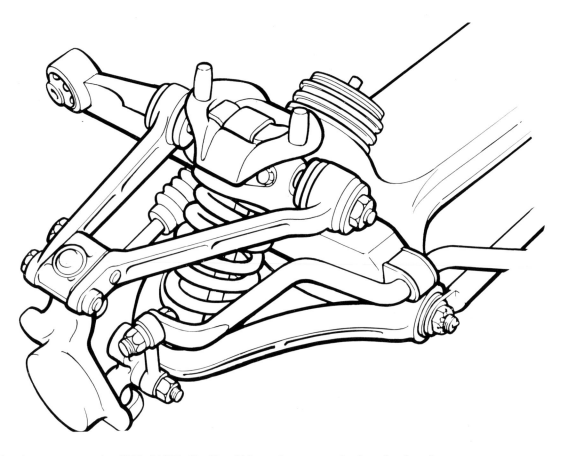

*For the new-generation XJ6 of 1986, Jim Randle's engineers completely redeveloped
the typical Jaguar coil spring/wishbone independent front suspension system,
but retained the separate subframe, for refinement purposes.*

high standards of refinement and comfort, but the XJ40 would have to be even better.

It was a daunting task. Everyone at Jaguar realized that the new XJ40 would have to be a real advance on the old Series III, and that to match Mercedes-Benz and BMW standards it would have to be better in all respects. Randle set targets for every area. He already knew that the new car would be at least as roomy as the Series III, but most notably he wanted his department to deliver better aerodynamics, better fuel efficiency, better handling, and no reduction in refinement, while improving the product's quality, and life expectancy. If

possible he also wanted it to cost less to build. This would have been difficult enough if the team was building on well-established components. But on this occasion Jaguar's engineers had a doubly-difficult task – they were designing a *completely* new model.

All-new cars are very rare indeed, especially in modern times when capital costs are so high. These days the world's motor industry talks cosily about engines, transmissions and even suspension assemblies as 'building blocks' – for it is a series of 'building blocks' which go towards making up the whole. Most new models use some carry-over 'building blocks' from old models, which

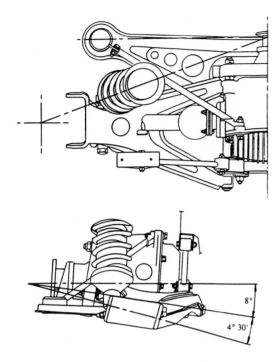

*For the XJ40, Jaguar developed a new type of independent rear suspension, with only one coil spring per side (instead of two on the original XJ6 types).*

helps reduce tooling costs, takes the pressure off at least one aspect of the design, and sometimes reduces the time needed to get a new car right. BMW's then-current 7-Series used engines and transmissions from the old models, Jaguar's Series III still used engines which had evolved in the 1940s, and even the sensational E-type of 1961 had used 1940s-type engines, transmissions and front suspension layouts. Not even in 1948, when the company had launched the XK120 and Mk V models, had there been an all-new Jaguar model, for the gearbox and axle were carry-over designs. So when Jim Randle's engineers set out to attempt the impossible – and, make no mistake, there were siren voices wailing that it *was* impossible – it was the first time Jaguar had ever taken so large a step.

XJ40, in other words, was to have all-new bodyshell, engine, transmission, front suspension and rear suspension assemblies. No major feature, and no major component, would be carried over from the Series III type. Not only did this mean more to get right, but more to get right *at the same time*. By the time XJ40 reached the showrooms, in fact, it was no longer all-new, because the engine and transmission units had already been blooded in the XJ-S model – but when it was conceived this had not been a factor.

## ENGINES AND GEARBOXES

Work on a new generation six-cylinder engine had begun, haltingly, in 1970, but the true ancestors of what became known as 'AJ6' were not conceived until 1976. Even after this it took a further seven years of development and refinement before the first production engines were fitted to the XJ-S 3.6-litre models.

In the early stages several different ways of developing a new engine were considered. In each case the idea was to produce a new engine without spending a fortune on tooling, and on new machinery: these projects are detailed in the separate panel.

For the 1980s, it was not only important to finalize a new engine layout, but to decide on sizes, specifications and power outputs. Jaguar's planners wanted to be able to use 'building blocks' which had smaller capacities, were lighter and more powerful than the old-type XK units. At about the time that the XJ6 Series III was launched, the BL board gave approval for the *next* family of engines to be tooled, allocating £32 million for this purpose. A dedicated area of the Radford plant was set aside for new machining lines to be installed – some to look after cylinder block work, some cylinder head work, and some crankshaft work – and a new family of six-cylinder engines evolved

*In the 1980s, Jaguar introduced the new AJ6 family of six-cylinder engines. The most powerful versions were those with this type of 24-valve twin-overhead-cam cylinder head.*

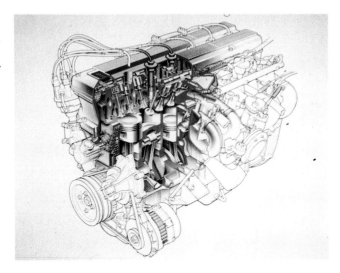

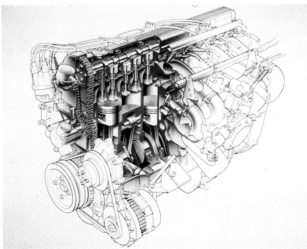

*The single-overhead-cam, 12-valve version of the AJ6 engine was only ever produced in 2.9-litre form, for use in the new-generation XJ6. In many ways, its head, camshaft and breathing layout was like that of the famous V12 engine, though the two heads were not interchangeable.*

| 4.2-litre XK engine (fuel-injected) | 3.6-litre AJ6 engine (1984 XJ-S type) |
|---|---|
| 6 cylinders | 6 cylinders |
| 4,235cc | 3,590cc |
| 92.05mm bore × 106mm stroke | 91mm bore × 92mm stroke |
| 2 OHC | 2 OHC |
| 2 valves per cylinder | 4 valves per cylinder |
| Cast iron block | Aluminium block |
| 205bhp (DIN) at 5,000rpm | 228bhp (DIN) at 5,300rpm |
| 236lb ft torque at 2,750rpm | 240lb ft torque at 4,000rpm |
| Weight 553lb (251kg) | Weight 430lb (195kg) |

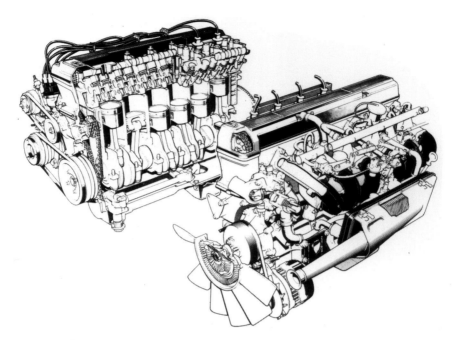

*Two cutaway views of the new AJ6 engine family, as originally used in the XJ40-type of XJ6. On the left is the 3.6-litre 24-valve twin-cam, while on the right is the 2.9-litre 12-valve single-cam design.*

### The AJ6 six-cylinder Engine

When the new AJ6 engine was unveiled in 1983, the idea of a replacement for the wonderful old XK engine had been kicking around at Browns Lane for thirteen years. After the V8 derivative of the V12 had been abandoned, no other layout than a straight six was ever seriously considered.

First thoughts were of effectively using half of the new V12 engine as a slant six, but this would only have been a 2.65-litre unit producing about 150bhp. A proposal to give this slant six a long stroke was also considered, but such a 3.4-litre engine could not then be machined on the Radford factory's facilities. It was time to think again. The idea of wedding a new 24-valve head to a modified (90 × 100mm) XK bottom end was tried, then rejected, and in 1976 it was decided that a completely new design, coded AJ6 should be developed instead. Harry Mundy led the project, but Trevor Crisp, who took over from him in 1980, was always part of the project team.

It was decided to produce 12-valve single-overhead-cam and 24-valve twin-overhead-cam versions of the new design. To give space around the engine, particularly for the inlet manifolding and fuel injection, the unit was designed to be canted over fifteen degrees to the near side of the engine bay. The XK had used a cast-iron cylinder block, but in this case an aluminium alloy block was chosen instead. The new engine, incidentally, had a cast rather than a forged crankshaft.

The first prototypes ran in 1979, using cogged belt drive to the camshafts and other auxiliaries, but duplex chains were substituted at a later stage, including chain drive to the power-steering pump. There were absolutely no common components with the old XK design, and any resemblance to the V12 unit was gradually lost as development proceeded.

The first AJ6 to be revealed was the 24-valve 3,590cc twin-cam of 1983 , which developed 225bhp when installed in the XJ-S 3.6, while the first (and, as it transpired, the short-lived) 12-valve single cam was the short-stroke 2,929cc/165bhp unit used in the new-generation XJ6 from 1986 to 1990.

Since then, of course, the enlargement and improvement of this engine has continued, and we may be sure that a lot more change will be seen in the 1990s.

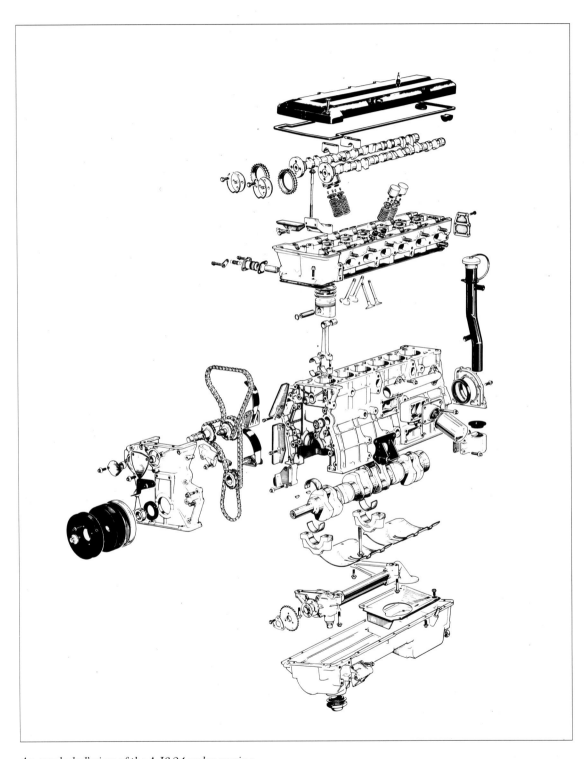

*An 'exploded' view of the AJ6 24-valve engine.*

to take best advantage of these. To meet the 'more power/lighter/smaller capacity' conundrum, the design team eventually settled on a swept volume of 3.6 litres, and it is instructive to compare the 'bare bones' of the new and the old types of twin-cams (*see* page 136).

The new engine, therefore, produced 11 per cent more power, but had a 15 per cent smaller swept volume, and was 21 per cent lighter. Although this was not stated when the new engine was launched in 1983, it could also be made larger, to look after future power and torque increases.

No sooner had the designers settled on the layout of the smart new engine than the sales force identified the need for a smaller and simpler derivative. As with the *original* XJ6 of 1968, there was a need for a smaller engine to satisfy some complex fiscal (insurance and taxation) 'break points' in European markets. There was an important break point at around 3.0 litres (to explain this fully, I would need a lawyer's training and a *lot* more space in which to bore the reader to death . . .), so to meet this the engineers produced a 2.9-litre version of the new engine, which was also fitted with a single overhead cylinder head. In many ways the new 2.9-litre was related to the V12 engine, for it had the same basic type of vertical valve single-cam layout, two valves per cylinder, the same Michael May 'Fireball' combustion chamber, and other details. It used the same cylinder block and cylinder bore dimension as the 3.6-litre unit, but had a shorter, 74.8mm, stroke. Once again, it is instructive to compare the new 2.9 with the last of the 3.4-litre XK engines as used in the Series III (*see* below).

As most engineers would have expected, the short-stroke single-cam 2.9-litre AJ6 was equally powerful, but was higher-revving than the old 3.4, and was also lacking in torque.

At this time, there was absolutely no intention to make a V12-engined version of the new car. The basic design of XJ40 was being finalized at a time when the second oil supply crisis (triggered by the Iranian revolution) had erupted, and when sales of existing V12-engined Jaguars had slumped alarmingly.

In the end, this decision proved to be costly for Jaguar. Once sales had risen sharply, and the XJ-S in particular had taken on a new lease of life, the V12 engine came back into favour, not only at Jaguar, but at rival concerns. Once it was learned that BMW and Mercedes-Benz were developing brand-new V12 models, Jaguar's management had to change its mind, and start looking at a V12 version once again. But by this time the XJ40 bodyshell, and all its panels, had been finalized, and was being tooled, and a major and expensive carve-up, with many new internal body panels, would be needed for the 1990s!

Choosing transmissions was equally

| 3.4-litre XK engine | 2.9-litre AJ6 engine |
| --- | --- |
| 6 cylinders | 6 cylinders |
| 3,442cc | 2,919cc |
| 83mm bore × 106mm stroke | 91mm bore × 74.8mm stroke |
| 2 OHC | 1 OHC |
| 2 valves per cylinder | 2 valves per cylinder |
| Cast iron block | Aluminium block |
| 160bhp (DIN) at 5,000rpm | 165bhp (DIN) at 5,600rpm |
| 189lb ft torque at 3,500rpm | 176lb ft torque at 4,000rpm |

The AJ6 (4.0-litre in this car) was almost as impressive to look at, and to stroke, as it was to provide a lot of very real power in 1990s-style XJ6s.

In 1991 you got a badge, but no name, on the front of the XJ6.

The inlet side of the early-1990s AJ6 engine was festooned with electronic measuring devices, fuel-metering gear, and anti-exhaust-emission fittings. By comparison with the 1960s, engines were definitely not to be maintained by DIY mechanics.

complex, for Jim Randle's team was still not happy with the general quality of the Rover-sourced '77mm' five-speed manual gearbox (as used in the Series III cars), and an early proposal to use a new design of Borg Warner automatic transmission, the Type 85, was upset when that project was cancelled. In the end, Jaguar decided not to struggle on with the improvement of the Rover five-speed gearbox, but instead it chose a five-speed transmission from Getrag of West Germany; the same Type 265 gearbox was used in the AJ6-engined XJ-S from 1983. Getrag was already a major supplier to BMW, so the appropriate high quality of the product could be assumed. Jaguar also assessed all manner of automatic transmissions in the next few years. Boxes from Ford, GM and ZF were all tried before the engineers settled on the latest ZF4 HP22 four-speed unit, a design which had also been chosen for use in the latest BMW 5-Series and 7-Series models.

*For the new generation of Jaguars, a new type of automatic transmission control and 'gate' was developed. For all the obvious publicity reasons it was known as the 'J-gate', but it also became known as the 'Randle-handle' to indicate Jim Randle's part in its concept.*

## TESTING, TESTING . . .

In the five years between the completion of the first prototype, and the start-up of series production, Jaguar's engineers completed the most intensive test programme they ever tackled. Early work was concentrated on what Jaguar calls 'simulators' – Series III types fitted with XJ40 components – after which five semi-engineered prototypes (SEPs) were produced in 1981 and 1982. To follow these cars, no fewer than twenty second-phase test cars – fully engineered prototypes (FEPs) – were built in 1982 and 1984. Some were running cars, some were test and development shells, and some were submitted to the indignities of a crash test programme. In the end, nearly ninety test cars were used in engineering work of one type or another!

One particular 'test car' was the 'buggy', which was merely a tubular steel chassis to which various XJ40 components could be fitted, and tested in *pavé* conditions. By the time XJ40 went public in 1986, this strange machine had completed well over 7,000 miles (11,250km), all on *pavé*, and its original frame (which was unrelated to the XJ40's structure itself) was still undamaged, and in one piece.

Early aerodynamic testing soon confirmed that the new car had a much better shape than the old – the drag coefficient (Cd) was well down, to a creditable 0.368 – and even though it had a larger cross-sectional area it was clear that it was going to be a significantly faster car than the old model.

Laboratory and rig testing is one thing, but because there is no substitute for hammering a new car on the roads in extreme conditions, XJ40 prototypes soon found themselves being thrashed all round the world. Naturally, disguised cars were tested in the UK (mostly at BL's Gaydon proving ground), and in Europe, with much ultra-high-speed work done at the circular Nardo track in Southern Italy. For more specialized work there was testing in Ontario, Canada, in the depths of winter, at Phoenix in the USA where the hottest climate could

# XJ6 ('XJ40 type' – introduced 1986)

**3.6-litre (1986–1989)**

## Layout
Unit-construction monocoque five-seater, front engine/rear-drive, sold as four-door saloon

## Engine
| | |
|---|---|
| Block material | Aluminium |
| Head material | Aluminium |
| Cylinders | 6 in-line |
| Cooling | Water |
| Bore and stroke | 91 × 92mm |
| Capacity | 3,590cc |
| Main bearings | 7 |
| Valves | 4 per cylinder; 2 OHC operation |
| Compression ratio | 9.6:1 |
| Fuel supply | Lucas electronic fuel injection |
| Max. power (DIN) | 221bhp @ 5,000rpm |
| Max. torque | 248lb/ft @ 4,000rpm |

## Transmission
| | |
|---|---|
| Clutch | Single dry plate; diaphragm spring; hydraulically operated |

## Internal gearbox ratios
| | |
|---|---|
| Top | 0.760:1 |
| 4th | 1.000:1 |
| 3rd | 1.391:1 |
| 2nd | 2.056:1 |
| 1st | 3.573:1 |
| Reverse | 3.46:1 |
| Final drive | 3.54:1 |

Optional ZF4 HP22 automatic transmission, with torque converter

## Internal ratios
| | |
|---|---|
| Top | 0.730:1 |
| 3rd | 1.000:1 |
| 2nd | 1.480:1 |
| 1st | 2.48:1 |
| Reverse | 2.09:1 |
| Maximum converter multiplication | 2.0:1 |
| Final drive | 3.54:1 |

## Suspension and steering
| | |
|---|---|
| Front | Independent by coil springs, wishbones, anti-roll bar and telescopic dampers |
| Rear | Independent, by coil springs, lower wishbones, fixed-length drive shafts, telescopic dampers |
| Steering | Rack-and-pinion, power-assisted |
| Tyres | 220/65VR390 |
| Wheels | Pressed steel disc (optional alloy, standard on Daimler) |
| Rim width | 7.0in |

## Brakes

| | |
|---|---|
| Type | Disc brakes at front and rear |
| Size | 11.6in diameter front discs, 10.9in rear discs, with vacuum servo assistance |

## Dimensions (in/mm)

| | |
|---|---|
| Track | |
|   Front | 59.1/1,501 |
|   Rear | 59/1,498 |
| Wheelbase | 113/2,870 |
| Overall length | 196.4/4,988 |
| Overall width | 78.9/2,005 |
| Overall height | 54.3/1,380 (53.5/1,358 on Sovereign/Daimler models) |
| Unladen weight | 3,903lb/1,770kg |

### 2.9-litre (1986–1990)

Specification as for 3.6-litre except for:

## Engine

| | |
|---|---|
| Bore and stroke | 91 × 74.8mm |
| Capacity | 2,919cc |
| Valves | 2 per cylinder; single OHC operation |
| Compression ratio | 12.6:1 |
| Fuel supply | Bosch electronic fuel injection |
| Max. power (DIN) | 165bhp @ 5,600rpm |
| Max. torque | 176lb/ft @ 4,000rpm |

## Transmission

| | |
|---|---|
| Final Drive | 3.77:1 (manual), 4.09:1 (automatic) |

## Dimensions

| | |
|---|---|
| Unladen weight | 3,793lb/1,720kg |

### 4.0-litre (Introduced 1989)

Specification as for original 3.6-litre except for:

## Engine

| | |
|---|---|
| Bore and stroke | 91 × 102mm |
| Capacity | 3,980cc |
| Compression ratio | 9.5:1 |
| Max. power (DIN) | 235bhp @ 4,750rpm |
| Max. torque | 285lb/ft @ 3,750rpm |

### 3.2-litre (Introduced 1990)

Specification as for original 2.9-litre except for:

## Engine

| | |
|---|---|
| Bore and stroke | 91 × 83mm |
| Capacity | 3,239cc |
| Valves | 4 per cylinder; 2 OHC operation |
| Compression ratio | 9.75:1 |
| Fuel supply | Lucas electronic fuel injection |
| Max. power (DIN) | 200bhp @ 5,250rpm |
| Max. torque | 220lb/ft @ 4,000rpm |

## Transmission

| | |
|---|---|
| Final Drive | 3.77:1 (manual), 4.09:1 (automatic) |

be found, in Australia for marathon runs in rough and dusty terrain, in Oman (heat, desert and glaring sunshine), and in the European mountains where one can find high passes for brake testing, thin air to puzzle the engine management systems, and high temperatures to make it more pleasant for the crews, and more stressful on the cars!

Well before the XJ40 had taken on its official name of 'XJ6', the design team was happy that it had solved most of the problems. By the time the first production cars were assembled at Browns Lane, endurance test cars had completed more than 5,500,000 miles (8,850,000km). By 1986 Jim Randle was quietly confident, if never totally satisfied. Everything possible that could be done, had been done. If the production cars were as good as hoped, they were bound to sell well.

## 'GIVE US THE TOOLS'

As Manufacturing Director, Mike Beasley had to prepare the business to produce a new model range without disrupting the flow of existing cars. In a gradual process which took five years, new facilities were installed at Browns Lane, at Radford and at Castle Bromwich, and when 'the button was pressed', assembly got smoothly under way.

The AJ6 engine was ready first, with major new transfer lines of machine tools being installed at Radford, and with small-scale production beginning in 1983 so that 3.6-litre 24-valve units could be fitted to XJ-S Cabriolets from that time. Three years later, only small changes were needed to this engine before it was made ready for use in the XJ40 saloon, but the first 'off-tools'

---

### The Castle Bromwich Factory

Jaguar's Castle Bromwich body plant has had a complex history, and has only been dedicated to Jaguar production since the 1980s. Until that time, Jaguar had never had its own body supply plant: ever since the Mk VII was launched in 1950, its saloon shells and monocoques had been sourced from the Pressed Steel Company at Cowley, near Oxford. Although the Castle Bromwich factory is only five miles north-east of the centre of Birmingham, alongside the A452 and very close to Junction 5 of the M6 Motorway, the site overlooked green fields until the 1940s.

The first buildings on this site were erected by the Nuffield Organization, as a shadow factory for extra aeroplane production. Building work began in 1938, and the plant was ready to start building Spitfire fighters by mid-1940. At first, the Nuffield Organization managed it for Vickers Armstong, but after a blazing row between Lord Nuffield and Lord Beaverbrook it was reallocated to Vickers.

During World War II the plant was expanded enormously, producing 12,000 Spitfires and more than 300 Lancasters, while the area on the other side of the main road became Castle Bromwich aerodrome. After the war, when military production was closed down, this massive plant was taken over by Fisher & Ludlow, then an independent concern, to produce car bodies. F & L was absorbed by BMC in 1953, and soon began to concentrate on supplying body structures to Longbridge and other BMC plants.

After the formation of British Leyland, what was still colloquially known as 'the Fisher factory' became a component of Pressed Steel Fisher, and by the end of the 1970s stamping and assembly of Jaguar bodyshells had been moved (from Cowley and Swindon) to this site.

By the time that Jaguar was privatized, in 1984, the Castle Bromwich factories were busying themselves solely with the assembly and painting of Jaguar monocoques. In recent years there has been further rationalization, for pressings are now supplied from the joint GKN-Jaguar factory at Telford, in Shropshire.

In the early 1990s there was a lot of spare factory space at Castle Bromwich, and it is an open secret that Jaguar might consider setting up new car assembly lines there if and when the smaller Jaguar is prepared for sale.

*The XJ-S was face-lifted in 1991, but few changes were obvious from the front, especially on the convertible model.*

single-cam 12-valve 2.9-litre units were not needed until that time. Then, as now, the AJ6 facilities were totally dedicated to this power unit, in their own location at Radford, for the big V12 units take shape in a different part of the factory, and the old-style XK engines are assembled a considerable distance away, on another part of the site. After each engine is completed, it is immediately mated to its transmission, and the assembly is then fixed to mobile test beds, transferred into test buildings, and 'hot tested' before being trucked the few miles across Coventry to the Browns Lane plant.

The contract to manufacture body press tools, and to supply panel sets, went to what most enthusiasts still know as Pressed Steel, though by the early 1980s that company was merely a part – but an important part – of the BL organization. Although the original XJ6 shells were assembled under Pressed Steel/Pressed Steel Fisher control at Castle Bromwich, near Birmingham, by the early 1980s Jaguar had taken over control of that rambling factory site. Right from the start, therefore, Jaguar decided to invest heavily in a brand-new body assembly facility at Castle Bromwich, the result being that new-style XJ6, old-style V12-engined Series III saloons *and* XJ-S bodies all came to be assembled there, in different factory blocks. Each type of shell subsequently went through the same modern paint plant before being transported to Browns Lane for final assembly to take place.

The XJ40 facility, a thoroughly modern layout of welding fixtures and robots, was installed in an empty factory block in 1984–5, and was dedicated solely to this type of shell, or derivatives of it. When I made a tour of this building (which used to make BMC bodyshells for cars like the Mini and the A40), my guide proudly pointed out how flexible the robotization was, and confirmed that it could cope with different versions of the same car. What different versions? At the time, his lips were sealed . . .

Over the years, Browns Lane had become less of a manufacturing plant, and more of an assembly factory. In the 1950s and early 1960s Jaguar had not only machined, assembled and tested its own engines and gearboxes on this site, along with all the trim and wood manufacture, but saw the bodies arriving 'in white', and had a paint shop in the south-west corner. All design, development and proving work was carried out in shops dotted around the site. By the 1980s all engine, transmission and running gear manufacture had been moved to Radford, leaving the much expanded trim, seating and wood manufacturing side of the business at Browns Lane.

Even so, this well-developed layout was itself completely reshuffled once again, this time to look after assembly and pre-delivery testing of the new XJ40. In the early 1980s, the main (oldest) Browns Lane building had housed assembly of Series III and XJ-S models, while Daimler Limousines were assembled in their own special department to one side of this building. In the meantime, pre-production XJ40s were built in a shop which once specialized in E-type assembly, and was later a paint shop. Engineering work was concentrated in the old GEC building on the south-west perimeter.

From mid-1986, however, Daimler Limousine assembly had been moved to another separate building at Browns Lane, the old 'E-type' shop, was given over to the assembly of V12-engined Series III saloons, while Browns Lane had been re-equipped to produce new style XJ40-type XJ6s, and the XJ-S.

Mike Beasley became Sir John Egan's assistant in 1986, after which Derek Waeland became Manufacturing Director. Once Sir John Egan had seen design, engineering and development moved to the newly purchased Whitley plant on the south of the city, this was the layout which Jaguar hoped would allow it to build up production, profits *and* reputation in the late 1980s and early 1990s. The 'squeeze point' – that which would limit production of cars if demand continued to rise – was in the paint shop at Castle Bromwich.

## XJ6 ON SALE – THE EARLY YEARS

Because the new model took so long to develop, and because so many test cars were used in so many parts of the world, it was impossible to keep the project under wraps until the last minute. Even by 1983, when the AJ6 engine made its sales début in the XJ-S Cabriolet, writers in motoring magazines were beginning to spread delicious rumours about a new saloon car – and were getting it wrong! In October 1983 Britain's *Autocar* staff wrote that 'AJ6 is of course the engine for XJ40, the replacement Jaguar saloon promised for the next motor show . . .'. That promise, of a new model in 1984, was soon made redundant, for spy pictures of the new car captioned with vaguer and vaguer launch dates appeared in future issues.

On the other hand, the same article included two other pertinent comments. 'It is also believed that the V12 will not fit XJ40 . . .

*The 'entry-level' new-generation XJ6 of 1986 had a 2.9-litre engine, and its recognition point, from this angle, was the use of twinned circular headlamps. The 12-valve OHC 2.9 was soon displaced by the more powerful 24-valve 2OHC 3.2-litre.*

the prestige value of an expensive, limited-production, refined fire-breathing saloon is invaluable, and would seem to be a marketing opportunity which should not be missed' – and – '. . . the 2.9-litre AJ6 could have a good future, but not, if our information is correct, in XJ40. So powered, that will be seriously under-engined.'

Jaguar had to contend with these rumours, but as its annual sales were continuously growing throughout the early 1980s, no serious attempt was made to plug the gaps in security. After all, if 18,455 Series IIIs could be sold in 1982, 26,668 in 1984, and no fewer than 30,218 in 1985, no one was delaying a purchase to wait for a new model!

In the meantime David Boole, who had masterminded the successful (and protracted) launch of BL's Mini Metro in 1980, had become Director of Communications and Public Affairs at Jaguar, and now set out to repeat the trick with the new Jaguar. After becoming project manager for the launch, one of his major problems was to convince the press, then the customers, that the new car was to be badged XJ6, and *not* XJ40 as it had been coded in private (and in public!) for so long. Helped along by David's expertise in the delicate business of drip-feeding information to influential sources, of building up expectations ahead of the official launch, and of pointing out how

*Even though it had been in production for nineteen years when the 1991 XJ-S was revealed, the V12 engine, particularly its architecture, still makes people gasp when they see it for the first time.*

*Every modern Jaguar has a carefully detailed fascia and instrument panel. The wood veneer, and the leather padding, is what customers expect of cars like these.*

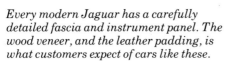

*Faced with restoring a new-generation XJ6 in the twenty-first century, this is the standard you will have to achieve with the door panel and 'furniture'.*

*Recognition points of the front corner of a four-headlamp XJ6 of 1991.*

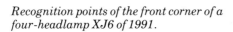

The face-lifted XJ-S of 1991 featured a choice of V12 or 4.0-litre AJ6 engines, and had different sheet metal and tail lamps at the rear.

Broad, heavy, and not very spacious for its two passengers, the XJ-S Convertible was often criticized as a 'wasteful' car — yet it is the most popular XJ-S type in several markets.

important the new car was to Jaguar's future, it was a great achievement that the cynical media, in particular, did not grow bored with waiting for an all-new Jaguar.

By the time XJ40 was launched, after all, the old-type XJ6 had been on sale for eighteen years. Yet when the public finally set eyes on the new machine, there was a chorus of praise, with pundits cooing over the style, engineers enthusing over the technical aspects of the car, and almost everyone appreciating the prices.

The launch finally came in October 1986, by which time the motoring, general and financial press, Jaguar dealers, financiers of the City of London, trade union leaders, politicians, famous personalities – in fact almost everyone who might be interested in seeing the new car, profiting from it, or buying one – had been exposed to its charms. In general, it received an enthusiastic response, particularly as it was not only seen as modern, and technically interesting, but as offering remarkable value for money.

Except that the V12-engined Series III saloon was continued, there was to be very little overlap between the old and the new. Because there had been time for Jaguar to build over 200 pre-production cars in 1985 and 1986 (not actually hand-building them, but at least taking time to see that everything fitted, and that everything worked properly), when series production began all the original range could be produced at once. At a stroke, therefore, Jaguar introduced five versions of the new car. Inside one basic four-door bodyshell there were two marque badges, two sizes of engine, a choice of manual or automatic transmission, and three different trim levels. With this in mind, it was easy to see why there was such a wide spread of prices. This is how the new range lined up (with the original late-1986 UK retail prices in brackets):

| | | |
|---|---|---|
| Jaguar XJ6 2.9-litre | 165bhp | (£16,495) |
| Jaguar XJ6 3.6-litre | 221bhp | (£22,995) |
| Jaguar Sovereign 2.9-litre | 165bhp | (£18,495) |
| Jaguar Sovereign 3.6-litre | 221bhp | (£24,995) |
| Daimler 3.6-litre | 221bhp | (£28,495) |

The entry-level 2.9-litre car was cunningly, and aggressively priced, for in the autumn of 1986 a 2.8-litre Ford Granada Scorpio cost £17,506, and a 2.5-litre Rover Sterling cost £18,795. It was prices like that – which were not being discounted in those stable days – which helped give the impression that the new Jaguar was such a bargain. And so it was – though, in fairness, I should record that the majority of customers bought the more comprehensively equipped Sovereign types!

As far as the car's performance, and its dynamic behaviour, was concerned, no one was complaining. The Series III chassis was a hard act to follow, but almost everyone seemed to agree that the new model was an advance in all respects. It looked good, it was roomy and refined, and it was as fast as expected. For such a large car, with such an overall executive character, it also handled remarkably well.

Few other car makers, in fact, could match Jaguar's unique contribution to the art of motor car development. It is one thing to provide a wonderfully supple ride – the ability to float over the potholes of city streets without transmitting crashes to the interior, the ability to remain stable at high cruising speeds – and it is quite another to provide prodigious quantities of grip in fast cornering. Jaguar provided both, neither being compromised by the other. It was only the unreconstructed 'sporting' Jaguar enthusiasts who wondered why the ride could not be tightened up even further.

There was a welcome for the car's comprehensive electrical and electronic equipment – no fewer than seven major microprocessors, and several smaller microprocessors, were fitted – and as ever there was praise for

*The twin-cam 24-valve 4-litre AJ6 engine was different in every way from the old-type XK engine, being more compact, lighter, and considerably more powerful. For installation reasons it was mounted at an angle in the engine bay of the XJ6.*

the interior of the cars, which combined leather, deep carpets and wood veneer in a typical Jaguar way. On the other hand, the most controversial aspect of the new car was its instrument panel layout, which featured large analogue speedometer and rev-counter dials surrounded by an electronic cluster of gauges and controls. Predictably, too, there were complaints that the 2.9-litre 12-valve-engined car was lacking in performance. Jaguar had never hidden the fact that the 2.9 was a tax-cheating special for certain markets, but by modern standards a top speed of 117mph (188kph) was somehow not enough for a £16,495 Jaguar especially as

the hard-worked engine tended to be *less* fuel-efficient than the 24-valve 3.6-litre type.

The customers, though, were not complaining. Once the all-new Jaguar saloon got into full production, and cars were delivered all around the world, sales boomed as never before. In 1985 – the last full year of old-style Series III production – 37,745 Jaguar cars had been sold, but in 1987 that figure surged up to 46,643, and in 1988 it reached an all-time high of 51,939. Nevertheless, re-touching of the range began soon after launch. Jaguar's planners listened carefully to their customers, their dealers,

*In 1991 this summarized Jaguar's three-body shape range.* Clockwise, from the top – XJ-S (Convertible, though Coupé also available), 'XJ40' type of 4.0-litre XJ6, and Series III V12.

and to the press, and set out to make the cars more powerful, and to return to a more traditional type of interior.

The first important up-date was announced in September 1989, just as the 100,000th example of the new-style car was produced. This saw a 4.0-litre 24-valve AJ6 engine taking over from the original 3.6-litre. The new engine had 235bhp compared with 221bhp but, more importantly, it had 285lb ft of torque compared with 249lb ft of torque; for British customers, a catalytic converter was offered as a £350 extra. At the same time there was a return to a more traditional style of instrument panel. Out went the digital and 'glowing light' displays, in their place came clear, easy-to-read, analogue dials instead.

A year later the Jaguar scene had changed once again, for the company had been taken over by Ford, and the 'first-evolution' process was completed with an engine change for the entry-level XJ6. Instead of the simple single-cam 12-valve 2.9-litre engine (which had not been a sales success, taking only 9 per cent of annual XJ6 sales) there was a newly developed twin-cam 24-valve 3.2-litre version of the AJ6 unit, which was effectively a short-stroke version of the 4.0-litre, and came with a three-way catalyst as standard. There was a lot more power and torque than before – 200bhp instead of 165bhp, and 220lb ft compared with 176lb ft – the result being a car which felt, and was, a lot faster than before.

The top speed rose from 117mph (188kph) to 135mph (217kph), and the fuel consumption actually improved slightly, from around 19mpg (14.9l/100km) to 20mpg (14.1l/100km). It was no wonder that *Autocar & Motor*'s testers described the car as 'a fine all-rounder and great value'.

At the same time Jaguar also introduced an optional 'sports pack' for all the cars, this featuring new forged alloy wheels and 225/55ZR rated tyres, lowered ride heights, stiffer springs and dampers, a firmer front anti-roll bar and steering with less hydraulic assistance.

For 1991, therefore, UK prices started at £23,750 for the four-headlamped XJ6, rising to £35,500 for the 4.0-litre Sovereign, and topping out at £39,500 for the Daimler 4.0-litre. Yet there was still no sign of a V12-engined version of this car, for Series III V12 assembly continued into the early 1990s. By that time, in fact, Jaguar's sales, and finances, had taken a pounding, for the British motor industry had slipped into deep recession. Happily, however, the company's future was underpinned by the might of Ford, which was determined to get a return from its hefty £1.6 billion of investment. In one form or another, the 'XJ40' generation of Jaguar saloons was set to be in production for several more years.

# 9 JaguarSport

In May 1988 a simple little announcement paved the way for an exciting extension of Jaguar's business:

Jaguar and the Tom Walkinshaw Racing Group are forming a new company to produce high-performance versions of the current Jaguar range.

The first product from JaguarSport Ltd, will be based on the XJ-S, and will be available later in the year . . .

It was the start of a project which is still growing, and which should result in the launch of more and yet more exciting derivatives of Jaguar models in the years to come. JaguarSport is a 50/50 joint venture between the two companies, which was launched at an initial cost of £5 million, with Sir John Egan as the company's Chairman, and Tom Walkinshaw as its Managing Director.

Walkinshaw originally stated that his principal aim in life was 'to produce limited volumes of uniquely styled, high-performance cars, which will help broaden the marque's appeal to customers who require their Jaguars to have more overt sporting characteristics'. The company's initial aim was to produce up to 500 cars every year.

Bill Donnelly, Sales and Marketing Manager of JaguarSport since the company was formed, told me about the origins of JaguarSport.

The seeds were very directly planted by the success of the Jaguar Group C racing programme, and even that could be traced back to the Group A Touring cars. I think it was the fact that the relationship between TWR and Jaguar had proved itself over a number of years. The most obvious route was to take some of that racing expertise and to build it into road cars.

It was not the first time, of course, that other concerns had proposed to build a series of modified Jaguar production cars, and not the first time Jaguar had been approached to approve a joint venture. Until TWR and Jaguar got together, however, none had the credibility to make it look possible.

The heritage, and the pride that runs through the marque and the people at Browns Lane, didn't lead them very comfortably towards a joint venture where someone else was modifying, or improving in some ways, or adapting their products. In that respect it was a tremendous vote of confidence in TWR, that they were taken on as partners.

One of JaguarSport's main objectives, the reason it was set up, was to begin changing the reputation and character of the cars. During the 1970s and 1980s the character of the production cars had changed. Most people accept that they gradually became more luxurious and more comfortable, but that they lost that sporting edge. At the same time, what a marketing department would call the 'customer profile' also changed. The average age of an XJ6 owner was fifty-two years, and that of an XJ-S owner forty-nine years. JaguarSport's target was to change all that, to claw back some of the drivers who had defected to BMW or Porsche, and to bring back some of the sporting elements which had been present in the 1960s.

## JaguarSport

JaguarSport models are developed from the mainstream Jaguar products, but have the following differences:

### XJR-S Coupé (1988–1989)

Mechanically similar to the 5.3-litre V12 XJ-S Coupé, but with a revised suspension and steering package, new wheels and tyres, and a body 'dress-up' kit

### XJR 3.6 saloon (1988–1990)

Mechanically similar to the 3.6-litre XJ6 saloon, but with a revised suspension and steering package, new wheels and tyres, and a body 'dress-up' kit

### XJR-S 6-litre Coupé (Introduced 1989)

This was the updated version of the original XJR-S, but with the following revised engine and chassis:

## Engine

| | |
|---|---|
| Bore and stroke | 90 × 78.5mm |
| Capacity | 5,993cc |
| Max. power (DIN) | 318bhp @ 5,500rpm |
| Max. torque | 362lb/ft @ 3,750rpm |

## Suspension and steering

| | |
|---|---|
| Tyres | 225/50ZR – 16in (Front) |
| | 245/55ZR – 16in (Rear) |
| Rim width | 8.0in |

### XJR 4.0 saloon (Introduced 1990)

This was an updated version of the original XJR saloon, but with the following revised engine and suspension:

## Engine

| | |
|---|---|
| Bore and stroke | 91 × 102mm |
| Capacity | 3,980cc |
| Compression ratio | 9.75:1 |
| Max. power (DIN) | 251bhp @ 5,250rpm |
| Max. torque | 278lb/ft @ 4,000rpm |

## Suspension and steering

| | |
|---|---|
| Tyres | 225/55ZR – 16in |
| Rim width | 8.0in |

In Bill Donnelly's words,

We wanted to make the cars more driver-orientated. We had a very specific customer in mind, who was younger than the traditional Jaguar owner, who wanted a car that could be driven hard, rather than enjoyed in the more luxurious sense.

JaguarSport, though part of the TWR Group of Companies, was set up to operate independently of the rest. Except that they shared the same site at Kidlington until the end of 1991, there was no functional link with TWR Racing. However, the 6.0-litre V12 clearly owes a lot to the racing engines developed by TWR for use in Group C and IMSA models.

*The original JaguarSport XJR-S had different wheels, suspension, and body style details, but there were no changes to the V12 engine. One of the amazingly successful TWR Group C cars is in the background.*

## XJR-S – THE FIRST JAGUARSPORT PRODUCT

In August 1988, only three months after JaguarSport had been set up, the first new model was announced. This car, named the XJR-S, based on the V12-engined XJ-S and priced at £38,500 (£5,600 higher than the standard product), was badged to mark the Le Mans victory. The first one hundred cars featured special paint and trim schemes, victory laurels engraved on the treadplates, and a plate showing the car's build number; logically enough these cars were known as Celebration Coupés. 'R' for racing? Who knows . . .

Although there had been no time to make changes to the running gear, the XJR-S featured suspension and steering modifications, new wheels and tyres, along with body styling and interior trim changes. That, in fact, was only the start, for work had already started on improving the engine. Completing the development, and getting the differences homologated, took a further twelve months, so it was not until September 1989 that the XJR-S 6.0-litre, complete with 318bhp version of the engine, was made available. At this time the Jaguar-Sport transformation was only available on the coupé style.

In those days the completion of the cars followed a complex sequence. Standard cars were built at Browns Lane, Coventry before being transported south, about forty miles, to JaguarSport at Kidlington. This

*The 6.0-litre V12-engined XJR-S model, developed by JaguarSport at Kidlington, near Oxford, was introduced in August 1989.*

was where all the mechanical and style changes were carried out and (after the 6.0-litre V12 engine was made ready) where the engine swop took place. Cars were then delivered direct to their customers.

By 1991 the system had been simplified, in that all the modified parts were fitted at Browns Lane when the cars were being assembled, though the cars were still carried to JaguarSport for final setting up of the electronic engine management system, and of the suspension and steering geometry. The cars were then taken back to Coventry to be fed into the company's normal delivery system.

At first, JaguarSport admitted to having a capacity to produce about forty cars a month each of two types – or about 900 cars a year all told – but thought that about 250 XJR-S types could be sold in the UK. To begin with, every car was sold in the UK. Right away

twenty-two Jaguar dealers signed up to take JaguarSport products as 'Sports Specialists'; because each had to make a considerable investment in training and equipment JaguarSport didn't want to spread its franchise too thinly. Then, from May 1990, the XJR-S 6.0-litre went on sale in the six major European markets for Jaguar – Germany, France, Belgium, Holland, Italy and Spain – and soon after that the JaguarSport franchise was thrown open to any UK dealer.

But that was not all, for several further developments were announced in 1991. First of all, JaguarSport began to build XJR-S conversions on the basis of the face-lifted XJ-S, which was by no means the simple reorganization that it sounds (there were several body style differences to be accommodated), but next it was decided to apply the same package of improvements to the XJ-S Convertible. The reason for that, most significantly, was that JaguarSport had decided to break into the promising and potentially large USA market. From late 1991, with a 6.0-litre engine developing 333bhp in full European-market catalyser form, with even better handling than before, and with a top speed of 158mph (254kph), it was an intriguing package on which to make that leap, and the effort seemed to be worth it. As Bill Donnelly told me in 1991, 'The market is big, the potential is enormous. The Coupé *and* the Convertible will be on sale in the USA by the spring of 1992.' Even though the world market for cars like this turned down sharply in 1990 and 1991, JaguarSport was hoping for USA sales of 400–500 cars a year.

## XJR SALOON

Only a matter of months after the XJR-S had gone on sale, JaguarSport was ready to start selling its own interpretation of the new-generation XJ6 saloon. In the same way that it saw younger drivers ready to buy a more sporty XJ-S, JaguarSport saw a demand for better handling, more individual-looking four-door Jaguar saloons as well.

The transformation came in two stages. The first offering, unveiled in October 1988, was the XJR 3.6, where the four-door saloon was treated to a package which included uprated and more sporting suspension and damper settings, steering changes, new wheels and tyres and a number of aerodynamic add-ons, including new body sills, bumper mouldings and a rear spoiler. Every car was equipped with automatic transmission and the price was £38,500.

Then, following Jaguar's introduction of the 4.0-litre engine to replace the 3.6-litre, JaguarSport evolved the XJR 4.0, this taking over from the original XJR 3.6, and making its bow early in 1990. The engine had been slightly improved, producing 251bhp (7 per cent more than the standard car), and was

*When a customer ordered a JaguarSport XJR 4.0-litre, he got a faster and better handling car than the standard product, without ceding any of the refinement and practicalities.*

backed by the latest dual-mode ZF four-speed automatic transmission. As with the original 3.6-litre type, it was *only* offered with Jaguar badging, for there was never any intention to market JaguarSport Daimlers. This, JaguarSport point out, would have gone contrary to all the traditions of Daimler motoring.

On the XJ6 saloon the JaguarSport styling package was so discreet as to be unobtrusive. At the front there was a different bumper with a deeper spoiler, the sills flared slightly at front and rear to match the squatter tyre profiles and Speedline wheels, and the rear spoiler on the boot lid was small, functional, and almost invisible. But for an extra £6,950 over the Sovereign, and a 16bhp power increase which added a mere 3mph to the top speed, was the customer getting value for money? JaguarSport thought so, and *Autocar & Motor* testers confirmed this by dubbing the XJR 4.0 'the best Jaguar saloon yet'.

The most important advance – an advance not only on the standard saloon but on the XJR 3.6 – was the better handling and steering response; I spent a few days in a JaguarSport-suspended car and discovered a drive which was so nimble and effortless that I positively looked forward to long journeys.

By 1991, the TWR Group had grown to include thirty-two different companies and showed an annual turnover of more than £100 million. JaguarSport, with Bill Hayden at its head and Tom Walkinshaw as its Managing Director, was already the largest company in the group, yet there seemed to be no end to its growth, or to its ambition. Having built more than twenty XJR-15 racing two-seaters, and with the mid-engined XJ220 'supercar' almost ready to go on sale, JaguarSport planned to settle into more spacious premises at Bloxham, near Banbury, by the end of 1991. When the first XJR-S was launched in 1988 Jaguar-Sport employed just fifteen people, but by

*Compared with the standard product, the JaguarSport XJR 4.0-litre had re-touched styling at front, rear and sides, with new wheels, better handling and steering, and an increase in performance. Every year hundreds of customers confirm that this is a popular package.*

*Posed at Silverstone in 1991 were three very exclusive cars in the Jaguar range. Clockwise from top: the TWR-built XJR12 which won the Le Mans 24-Hour race in 1990, the XJR 4.0-litre saloon, and the XJR-S 6.0-litre V12.*

the end of 1991 the workforce had grown to more than one hundred – and that figure was increasing all the time.

An indulgence, or merely a flagship for Jaguar? Certainly it is a flagship – Jaguar-Sport has always consciously aimed for the top end of the range, and the ultimate in performance (the startlingly fast XJ220 confirms that) – but it is certainly not an indulgence. As Bill Donnelly reminded me, JaguarSport was always intended to be a commercially viable company – in layman's terms it was always intended to make a profit – and as *everyone* in the motor racing industry knows, it seems to be against Tom Walkinshaw's principles to tackle any project unless it shows a profit at the end of the line.

What happens next? For the moment JaguarSport is not telling, although I was reminded that it had not done more than glance at the idea of an XJR-S with the six-cylinder engine, and that the new Bloxham premises would need something to occupy itself as building activity of XJ220 two-seaters begins to wind down. JaguarSport, after all, is already a 'centre of excellence',

for producing high-performance cars, and it intends to stay that way in the future. As far as production cars are concerned, the good news is that JaguarSport is now an accepted member of the Jaguar family, and its needs are already being considered as Jaguar (under Ford control) develops its forward model plan. Even so, I wondered, with environmental pressures crowding in all around us, would there still be a place for JaguarSport in five or ten years time? Bill Donnelly thought there would:

I think there will always be room for people to express themselves with a motor car. I think that what we're selling here is not just a vehicle to get you from A to B. I don't think JaguarSport cars are bought as toys, I think they are all working cars. I think that all of the cars we sell are well used, and some of them do very high mileages.

But by the mid-1990s, what then? A hotter-than-usual V12-engined four-door saloon? Four-wheel-drive conversions? A smaller-engined supercar?

# 10  Testers' Opinions

## XK-ENGINED XJ6

When the original XJ6 appeared, it was received with enthusiasm. In its introduction to a 1969 test *Autocar* thought that the 4.2-litre XJ6 'Sets new standards of ride, handling, quietness and refinement' and noted 'Easy 100mph plus cruising. Balanced cornering and brakes. Superb adhesion, unbelievable value. Best there is.'

Which meant that further study of the text was really not necessary, except to find this succinct comment:

*Testers loved the XJ6 for its style, its refinement, and its handling. The 4.2-litre version was a fast car, and praised as such, but 2.8-litre cars (like this one) were not lively enough, and Jaguar Cars never provided them for assessment.*

If Jaguar were to double the price of the XJ6 and bill it as the best car in the world, we would be right there behind them ... As it stands at the moment, dynamically, it has no equal regardless of price, which explains those twelve-month delivery quotes from dealers ... We of *Autocar* set it as a new yardstick, a tremendous advance guaranteed to put it ahead for several years at least.

On this automatic transmission car, the top speed was 120mph (193kph), with 0–100mph (160kph) in 30.1sec, but with 15.2mpg (18.6l/100km) as the overall fuel consumption figure. For Jaguar, the only worrying comment must have been that: 'With stricter quality control (our test car developed two faulty door locks and a faulty wiper mechanism) and more expensive fittings it would wipe up the quality car market'. Even at that stage, it seems, the XJ6's early quality problems were becoming evident.

*Motor* was no less complimentary about its manual transmission car, heading its test 'British and Best', and stating that the 'Latest Jaguar sets impressive new standards; combination of performance, comfort, roadholding and quietness unrivalled at price; very few faults.' In the main test there were these comments:

The only qualms we have now about publishing this report is that it is going to make the waiting list even longer and Jaguar's problem of supply and demand even more acute. If you're on the list already and getting impatient, don't give up. It is worth the wait ... Jaguar have produced results which we believe every competitor throughout the world, from Rolls-Royce downwards, cannot afford to ignore ... in practically every department the XJ6 excels.

*Motor*'s car achieved 124mph (199kph) and 0–100mph (160kph) in 24.1sec, better

figures than those recorded for the automatic transmission version, and confirmed the power losses in the Borg Warner gearbox.

John Bolster of *Autosport* described the car as having:

High-speed silence and fantastic roadholding ... Boy! Bring me a large box of superlatives! I have just been testing a four-door saloon that corners better than a sports car, is more silent than a limousine, and costs less than half its true value! ... All Jaguars have been good lookers, but this one hits the jackpot ... As soon as one drives off, the uncanny silence of the car is noticed ... Even when driven hard, it is still unnaturally quiet ... Vivid acceleration is taken for granted in a Jaguar, but the cornering power at once astonishes ... The Jaguar XJ6 lives up to its advanced specification, and realization is even better than anticipation. No car is worthy of higher praise.

No British magazine ever tested the 2.8-litre car (Jaguar was well aware of its limited performance, and wasn't about to broadcast it to the world), but the 3.4-litre model, new in 1975, was provided for test. In the main, it was treated as the entry-level model which it was, for testers thought the performance and economy were both disappointing. *Autocar* thought it was 'a much more successful exercise than was the old 2.8-litre unit ... It does not offer the performance of the 4.2-litre, but it shows a worthwhile gain in economy, not to mention first cost'.

*Autosport* headlined its test 'A four-door XJ means value for money', and did its very best to be enthusiastic about the character of the 3.4-litre model although John Bolster thought the car was over-geared and commented that: 'At first, the car appears to lack performance, and of course its extra 4cwt [448lb/203kg] precludes Mk II acceleration. Nevertheless the stopwatch reveals that the figures are very respectable ...'

*Autocar*'s update on the 4.2-litre XJ6, in

Series III form, concentrated on the changes to the engine, and to the style:

Where Jaguar have made a really significant advance is in engine efficiency ... morning starts are virtually instant; also the warming-up driving period is utterly hesitation-free ... performance has been transformed, particularly at the top end ... we were particularly impressed by the way the AE Econocruise kept the desired speed ... an overall consumption figure of 16.8mpg [16.8l/100km] together with the very real performance gains shows how much Jaguar has increased the car's efficiency ...

All in all, the testers liked the design even more than they had in the past.

*Motor* confirmed these findings, particularly regarding better efficiency, describing the car as having 'More grace, space and pace', recording a 128mph (206kph) top speed, and pointing out the possibility of 20mpg [14.15l/100km] fuel consumption. The automatic transmission, however, came in for a great deal of criticism, not only for its refusal to kick-down into first gear above 30mph (50kph) (when instant response was wanted in heavy traffic), and because the car felt sluggish when exiting roundabouts.

## V12-ENGINED SALOONS

There had been such a long delay, and build-up of rumour (four years) that the press couldn't wait to get its hands on the XJ12 model. When *Autocar* sampled the new car, its enthusiasm gushed all over the pages: 'Phenomenal performance and 146mph [235kph] top speed, but deplorable fuel consumption. Superb quietness and refinement. Steering very light, and some sponginess in central position, but good directional stability and cornering ... A truly outstanding car.'

The 'deplorable fuel consumption' was a miserable 11.4mpg (24.8l/100km) overall, which was easily the worst recorded by any test car the magazine had 'figured' in recent years. There was a lot more, concerning the magnificent engine:

In terms of mechanical quietness, the XJ12 represents the nearest approach yet achieved to a car in which the only sensation of having a propulsive unit under the bonnet is that of speed and acceleration. It is not only the exceptionally low noise level, but the complete absence of any vibration or harshness as well, which makes the car so fantastically docile and effortless.

It was no wonder that the summary included this sentence: 'It is a marvellous achievement, deservedly the envy of the world.'

*Motor* actually tested a Daimler-badged version, the Double-Six, and recorded slightly lower performance, but confirmed all the major findings. Even so, there were pertinent comments about the steering and the failings of the transmission:

We are bound to record our belief that the car would be far easier and more pleasant to drive if its power steering had the proper resistance and feel characteristics ... What we dislike is the excessive lightness of the steering which calls for extra concentration at high speeds in cross-winds and makes it difficult to aim the car accurately in corners, let alone to sense reduced adhesion on slippery patches ...

Since *Autosport* had clearly loved the original XJ6, its reception for the XJ12 was predictable. Headlining his test 'Irresistible', John Bolster took it to France, and found that:

... the trip was a memorable experience. Most cars behave disgracefully on the more difficult roads of Northern France, but the Jaguar rode with that splendid disdain of

*The media always respected the XJ-S Coupé, but usually went overboard in praising the Convertible.*

*The new-generation XJ6's problem was that it had been rumoured for so long that some testers suffered from déjà vu when they finally got to try one. Everyone agreed that it was an advance on the old, particularly in feel and fuel efficiency.*

*The XJ-S HE enjoyed a friendly reception, not only for its better fuel efficiency, but for the signs of better build quality which went with it.*

*Jaguar was lucky that the new-generation XJ6 was introduced when the company's reputation was on the crest of a wave. Every published road test commented on the excellent overall package of styling, packaging, and dynamic behaviour.*

*The restyled XJ-S of 1991 was something of a let-down for those who had been expecting more changes.*

*Everyone seemed to love the XJ-S Convertible, especially when the sun was shining . . .*

bumps and pot-holes which has been almost a French monopoly . . . the most astonishing thing about the XJ12 is the way in which its acceleration persists at speeds where other fast cars are beginning to struggle.

But even John thought the fuel consumption appalling for 'the XJ12 only does about 14mpg [20.2l/100km] when creeping along like a snail at 70mph [113kph]'. Even so: 'This car is in many ways unapproachable and for a combination of smooth perform-ance, impeccable roadholding and luxurious ride it has no competitor . . . to the man with the right sort of money it must be just about irresistible.'

## TWO-DOOR COUPÉS

As already explained, the two-door cars had similar performance to the four-door types, so attention concentrated on their body lay-outs. When writing about the XJ5.3C in 1975, *Autocar* commented about the window sealing problem that 'it cannot be said that the problem has been completely solved since there is certainly some wind noise, but it is low enough not to be in any way annoying and at speeds under 60mph [100kph] is hardly discernible at all.' That sounds suspiciously like damning with faint praise to me . . . On the other hand, the comfort of the front seats came in for severe criticism, though the amount of rear space was thought 'most comfortable, and the amount of head, shoulder and kneeroom is generous by any standard.'

*Motor* thought that the XJ5.3C (complete with 285bhp fuel-injected engine) was 'One of Leyland's world-beaters. Effortless but startling performance from uncannily smooth and quiet V12 engine.' Regarding the body:

Compared to a 2+2 coupé, the Jaguar XJC is very spacious, being based on the original short-wheelbase saloon. There is ample leg room in the back . . . but there is a shortage of headroom for taller people . . . the Coupé is a comfortable four-seater. It is this extra space that caused more than one member of staff to query why Jaguar produce both this model and the XJ-S, which is much more cramped . . . most of us would go for the added interior room of the XJC.

## XJ-S

Since testers, to a man, loved the sexy over-indulgence of the E-types, one might have expected them to be supercritical of the XJ-S Coupés which took over in 1976, but this did not happen. In 1976, *Autocar*'s testers called the XJ-S a 'great car . . . very accelerative, fantastically flexible, highly refined – and in spite of its high price still cheaper than its international competition. A joy to drive.' In addition: 'the Jaguar XJ-S grand touring car is a reminder that we do still build cars that are world beaters . . . In many respects it is outstanding. In its combination of perform-ance with docility and refinement it is unapproached.'

The same article described the engine as 'extraordinary' with 'an imposing sound, hard to describe, rising excitedly but not loudly, the voice of great, continuing acceleration kept down, never raucous, even at the top end . . . there is not the slightest suggestion of a flat spot throughout its truly exceptional range . . .' and more in the same vein.

On the other hand, the car's considerable thirst – 15.4mpg (18.4l/100km) overall, but less than 10mpg (28.3l/100km) in the max-imum speed phase of the testing – was detailed, as was the lack of space in the rear seat: 'In the back the space is only occasio-nal, except for children. A 6ft adult will request those in the front to move forward for the sake of his legs and head, which must be stooped to clear the roof . . .' At this stage it was a test which glowed with superlatives, such as in the last sentence: 'We envy those who can find a place for this most covetable car.' Problems of product quality were still ahead of this model.

*Motor* was not quite as euphoric. After posing the question 'Tank or Supercar?', it answered itself with 'A bit of both. It's large, heavy, thirsty and cramped in the back. It's also superbly engineered, sensationally quick, very refined and magnificent to drive – a combination of qualities that no other car we've driven can match at the price.' In addition:

The Jaguar XJ-S is a magnificent car, not just for what it does, but for the way it does it. The XJ-S combines a startling performance with exceptional smoothness and tractability and a standard of refinement that few cars can match . . . none (other Jaguars apart) are smoother, quieter or more flexible.

*Motor* commented on visibility problems thus:

We expected the rear three-quarter pillars and curved Dino-style fins to create blind spots, but they are far enough back not to be a problem. Yet the windscreen pillars get under way (there are slender quarter window pillars and door pillars as well to clutter the side view) and the windscreen header rail is low and far back, giving a beetle-browed impression and a slight feeling of claustrophobia.

The testers, however, were reacting to post-energy crisis trends when they wrote that: 'The only real criticism that could be levelled at the XJ-S is that it's dated in concept. Like Concorde it is a superb technological achievement with perhaps a questionable future . . . the Jaguar is expensive, but still remarkable value for money.'

*Autosport*, naturally, went overboard about the car, describing it as 'nothing but the best' and 'there is still a rising demand for supercars and this, by far the most elaborate and luxurious Jaguar ever sold, is assured of a warm welcome.' Of the performance:

Really to enjoy this V12, it's best to have the manual transmission. Only then can one appreciate the steam-like flexibility of this wonderful engine. Where this one breaks new ground is in combining such flexibility with outstanding performance at high revs . . . The whole point of owning an XJ-S is that it is entirely unstressed at any normal speed, and stability which has been worked out for 150mph [241kph] is incomparably better than that of ordinary cars. I've driven the German and Italian equivalents, and very nice they are too, but nobody who has tried them all could be in any possible doubt. The V12 from Coventry is the one, and if they can make them all as well as they built the car they lent me, Britain has a world-beater.

By the time the HE derivative came along, the XJ-S's reputation had plumbed the depths, and was on the rise again. *Autocar* tested the latest type in 1982, placarding it 'Fireball efficiency', and summarizing that:

For all-round satisfaction – a tiger of a car when need be, the most refined, relatively speaking, of Grand Touring 2+2s when driven less hard, let alone gently, and always a gentleman in character – the Jaguar is even more so a paragon, especially now that its fuel economy is so much better than before.

(Overleaf) *Sales of the XJ-S Convertible were badly hit by the recession in luxury car sales in the USA, even though independent tests confirmed its qualities.*

The 'Fireball' cylinder head and the related changes had delivered a big improvement in fuel economy. The 'Economy' section of the test posed the question: 'The 20mpg XJ-S?', then answered it by:

No, not if you drive it as it begs to be driven when mood and road permit, fast – but yes, even the 21mpg [13.5l/100km] XJ-S if you pay only flippant regard to the British 70mph [113kph] speed limit . . . and provided you match 80mph [129kph] restraint with not more than half-throttle acceleration . . . the consumption varying between 20.1mpg [14l/100km] and a best interval of 21.2mpg [13.3l/100km].

Even so, this £19,708 test car averaged 16mpg [17.7l/100km], which was less than 1mpg better than before. Yet it was still a colossally fast machine, with a top speed of 153mph (246kph) 0–100mph (160kph) in a mere 15.7sec, and the ability to reach 130mph (209kph) in only 33.2sec. By almost any standards it stood at the top of its class – faster and more accelerative than £25,250 rivals such as the Porsche 928S, yet more economical than dinosaurs like the Ferrari 400i saloon or the Aston Martin V8.

*Autosport* thought that the XJ-S HE was 'even more of a Jaguar than any previous model . . . Perhaps no car is less tiring when driven on a long journey.' This time, too, it was happier with the fuel consumption for: 'Driven continuously at 140mph [225kph] it still cannot be called economical, but at more practical speeds it uses no more petrol than some cars of about half its size.'

When the new AJ6 engine was matched to the bulky XJ-S shell, the result was a car considerably slower than the V12-engined car, and although it was a cheaper car the testers were not as happy about the result. Of the 3.6-litre-engined Coupé, *Autocar* thought that the new car had below average all-round visibility, but was pleased by the performance – the top speed was a creditable 141mph (227kph), 0–100mph (160kph) came up in 19.7sec, and the team returned 17.6mpg (16l/100km) overall – the car obviously being more economical than the V12, but surprisingly fast.

When *Autocar & Motor* tried the 1991 model, in restyled 4-litre Coupé form, it was disappointed in the performance of what was still a very 'tight' test car. Recording a top speed of 136mph (219kph), and 0–60mph (100kph) in 8.7sec (the acceleration was something which smaller-engined hot hatchbacks could match, and was actually no better than for the 3.6-litre-engined car of the mid-1980s), testers wrote that the car was '. . . more charming than it was handsome – a gifted grotesque . . . The changes have made it look smarter from most angles.'

One comment was that:

It's disappointing enough that the Jaguar finds itself at the bottom of the performance league as defined by our chosen rivals . . . it fuels the trend for new Jaguars to be slower than their ostensibly less powerful predecessors. The catalytic converters are clearly taking their toll.

The balance, in fact, was in enjoyment rather than in performance:

. . . the big 4.0-litre engine doesn't give the big coupé quite the kick it needs but the unit has bags of character, works well with the new ZF four-speeder [automatic transmission], and is decently smooth and refined. An always capable chassis has benefited from further development and now provides genuinely enjoyable handling. And the new cabin is little short of a revelation. Head-jerking performance isn't part of the new XJ-S 4.0's repertoire, but it still has the power to seduce.

That last word was aptly chosen, for it was in the art of seduction which a Jaguar has always excelled. In 1991, as in 1986, the 'Big

Cat' still had a lot going for it.

The full Convertible, of course, filled a gap which testers had been nagging about for years. *Autocar & Motor* thought that it was

... a very complete and accomplished tourer in the true sense of the word. It is capable of taking two people and their luggage over long distances in great comfort. At £36,000 it may be the most expensive production Jaguar to date, but we feel it is worth every penny.

In Convertible form the V12-engined car had a top speed of 144mph (232kph), 0–100mph (160kph) acceleration in 20.4sec, and overall fuel consumption of 13.8mpg (20.5l/100km). There was no way, it seemed, that the big V12 engine was ever going to be a fuel-efficient unit.

*Autosport*'s view of the V12-engined convertible was that it was

... truly a thing of beauty. Clean, elegant and sleek, it is arguably the most stylish Jaguar around. Driving this luxury machine is the ultimate in self-indulgence. With its creamy smooth flow of power, there is total effortlessness in everything the car does ...

You can waft at remarkably high speeds with the top down: even at 100mph [160kph] there is little buffeting if you keep the side windows raised, up to 80mph [130kph] you don't have to raise your voice very much either ... Top up, it's fully weather-proof, and if not as totally refined as the coupé, it is highly civilized.

## NEW-GENERATION XJ6

Because it had been kept waiting for such a long time, you might have excused the motoring press for being tetchy about the XJ40 style of Jaguar saloon, but apart from an obvious disappointment over the performance of the 2.9-litre car, the new range got a very good reception.

*Autocar* was so impressed by the original 3.6-litre car that it headlined its test: 'Setting the cat among the pigeons', and went on to comment that the new car was '. . . faster, more economical and handles better and is well equipped to fight off German opposition . . .' The new car was clearly quicker than the old Series III, with a top speed of 137mph (220kph) (against 131mph/211kph), and 0–100mph (160kph) acceleration in 20.6sec (against 26.5sec). Equally impressive was the gain in fuel efficiency – 20.7mpg (13.7l/100km), which was 13 per cent better than before.

'Judged more subjectively, the manual 3.6 is a much more entertaining car to drive, quickly than its forerunner . . .' while refinement standards clearly remained high:

... in a lesser car, you undoubtedly wouldn't be able to hear some of the things which it is possible to detect in the Jaguar simply because the interior noise levels are *so* low ... although bump thump was just detectable over some of the tortuous roads of highland Scotland it is on the whole extremely well suppressed ... the damping provides a near magical ride/handling compromise ... the car feels far smaller than it is, and very nimble.

The verdict, overall, was that the new car offered 'tremendous value for money' and '. . . it becomes clear that Jaguar's biggest problem with the new car will be meeting the demand.'

Mike McCarthy drove a Sovereign 3.6 for *Autosport* and felt let down by the style:

I am one of those who is, frankly, disappointed with the shape. Now that you do see them on the roads there is blandness to them that renders them all but invisible. To me, a new Jaguar should turn heads ...

Even so, the new car's rivals were dismissed with:

But then, of course, they wouldn't be able to match the Jaguar's utter smoothness, the sensation of being wafted forward by a huge but gentle hand. At low speeds the AJ6 unit is all but silent, and when you accelerate hard all you hear is a muted purr . . .

McCarthy loved the chassis:

The other area where the XJ6 shines, of course, is in its ride/handling compromise . . . It will corner cleanly, with little roll, in a beautifully balanced way, and you really have to be trying hard – or doing something silly to upset its poise.

Mike also confirmed other opinions about the instrument and control layout: 'The detailing of the interior is, to me, the biggest disappointment of the car. The instruments, too, offend: the combination of analogue and digital just doesn't work . . .'

*Autocar* was polite, but little more, about the 2.9-litre engined car, pointing out that the acceleration was 'respectable' or 'acceptable', but that 'Other cars in its class certainly go faster'. The end-of-test summary made its point rather well:

The bottom line is that the 2.9-litre car is almost 1.5mpg less economical, at 19.3mpg [14.7l/100km] overall, showing that the engine needs to be worked much harder to maintain rapid progress. But the extra 20mph on top speed and 2.5 seconds off the 0–60mph time will set you back exactly £2,000 more. If you don't want that extra performance, then the XJ6 2.9 could well be the car for you.

When the up-engined versions came along, the verdict was even more favourable. Of the 4-litre, *Autocar & Motor* wrote that: 'In the past we've said the XJ6 is a great car in need of some thoughtful revisions. This car delivers them.' Not only was it faster than before, but it had the revised instrument panel where: 'The gimmicky electronics have been replaced by a conventional pack of gauges, and the trip computer has been simplified too. Jaguar listens . . . the 4-litre's full analogue set is a great improvement.'

In summary: 'The XJ6 is unquestionably a great car. It offers an unmatched blend of superb dynamics, high refinement and luxury ambience . . . we feel that this latest XJ6 is the car which should have appeared at the launch in 1986 . . . it is the best XJ6 yet.'

The 'Sports pack' equipped 3.2-litre model of 1990 was hailed as a 'Real Jag. at Ford Scorpio price', and as a '. . . fine all-rounder and great value . . . Jaguar has bent over backwards to ensure that it advanced the XJ6's cause rather than diluting it.'

The revised suspension was judged to provide '. . . a much more wieldy and agile car than the fluid handling but too softly set-up regular product – one which can be hustled along demanding roads with a feeling of greater security.' Most importantly: 'Judged as a replacement for the 2.9, the 3.2 is a knock-out success. It blows its predecessor away in every conceivable respect . . . we're more than happy to give the new entry-level XJ6 an enthusiastic recommendation.'

In 1990, as in 1968, the current XJ6 was an attractive and compelling proposition. There would certainly be more to come in future years.

# 11 Parts and Clubs

Once the worldwide 'Classic Car' movement had sprung up in the 1970s, almost every Jaguar model, whether new or old, whether numerous or rare, became popular and desirable. The demand for parts, and for them to be stocked long after the usual time limits had passed, continued to grow. No sooner had Jaguar started to run down its stocks of parts for a particular car than owners were pleading for them to be put back on sale. If such pleas were refused, a growing network of specialist workshops began investigating the manufacture of new parts. The result was that by the 1980s there was a huge 'cottage industry' which existed to keep older Jaguars on the road. At first this sort of expertise was concentrated on

When restoring an XJ6, don't dismiss this sort of detail as unimportant – always make sure that your restored car is correct in every detail. This is a Series I six-cylinder hub cap, while . . .

. . . this was the Double-Six equivalent in the mid-1970s.

In the 1990s, the cheapest Jaguars are those which are least fashionable. Pundits shun 2.8-litre XJ6s because they are not as fast as the 4.2-litre models, and because the engines had well-publicized problems when new. The surviving cars should be good bets in the 'classic' 1990s.

glamorous cars like the XK and E-type sports cars, but the 1960s generation of Mk IIs were soon swept into the net, and eventually the XJ models – saloons and coupés – followed.

As far as the cars covered in this book are concerned, it is almost always possible to restore them to full health. There are two very important reasons for this. One is that the cars themselves had a very long and successful life (which meant that what the motor industry calls the 'car park' was very large), and the other is that Jaguar itself has a policy of supporting the repair and renovation of all models until they have been out of production for at least ten years.

## PARTS SUPPLY

By the end of the 1980s, the original-style Series I and Series II XJ saloon models had been out of production for a decade, which meant that supplies of body panels was rapidly running down. 'Jaguar'-related items were much more freely available than equivalent 'Daimler' items. Trim and decoration items were almost extinct. Series I panels, where unique, had almost all disappeared, although Series II model panels were more readily available. Because a great deal of the Series III bodyshell (particularly above the waistline) was different from the earlier types, it was not possible to use those as 'carry-over' items, though many such panels could easily be reworked to include the correct piercings and mounting holes. Two-door panels were already scarce, and likely to become rapidly more so. Because the XJ40 type of saloon was still in production, the only obstacle to its repair and maintenance was the depth of the owner's pocket.

The situation regarding the XJ-S shell was much more encouraging than for the saloons. Except for details of badging and decoration, the coupé shell was not changed

---

**Daimler Derivatives of Jaguar Models**

Jaguar took over Daimler in 1960, and by the late 1960s had completely integrated the two businesses. After the Jaguar XJ6 was announced, it was almost inevitable that there would be Daimler equivalents of each model.

Here is the way that you can relate the two marques:

| Jaguar | Daimler |
|--------|---------|
| XJ6 | Sovereign |
| – | Vanden Plas 4.2 |
| XJ6C | Sovereign Coupé |
| XJ12 | Double-Six |
| – | Vanden Plas 5.3 |
| XJ12C | Double-Six Coupé |

From late 1983:

| | |
|--|--|
| Sovereign 4.2 | 4.2 |
| Sovereign HE | Double-Six |

Note: There has never been a Daimler version of the XJ-S model.

---

from 1975 to 1991, which meant that all panels were freely available. In the 1990s, the first body panel supply crisis was likely to concern the unique areas of the short-lived Cabriolet, while the supply of rear-end items of the full Convertible (built 1988–91) was also likely to be very restricted.

In general, Jaguar specialists reported no potential problems in procuring important mechanical parts for some years to come, though items unique to earlier types (as an example – carburettor-related fittings on V12 models, which went over to fuel injection in 1975) were likely to be the first to go out of stock. For those owners who are not sticklers for originality, in many cases it was possible to fit more modern parts to an earlier car.

The good news for enthusiasts, of course, was that there were thriving one-make clubs in many countries, all of which could advise on restoration and maintenance. Details of the most important of these are given below.

## CLUBS

For many years there was only one British-based club which catered for Jaguar cars and Jaguar enthusiasts. This had the blessing and help of the Jaguar factory itself. Then, early in the 1980s, all kinds of internecine strife broke out, many influential members left to start rival organizations, the factory took a step back from the raging dissent – and the result was that by 1990 there were three different clubs, with different aims and ambitions. It would be quite wrong for me to try to choose a 'Best Buy' from these clubs, so here is a summary of what they are, and how they look after the interests of Jaguar owners.

### Jaguar Drivers' Club Ltd

This is the oldest of the clubs, which has had a very controversial history in recent years. Top officials have resigned, others have rushed to take their place, and the HQ has been involved in legal action concerning alleged financial irregularities.

The original JDC was set up by Jaguar, as a descendant of the original SS Car Club of 1932, but eventually became independent of the factory. It grew to offer many services, not least of them being technical advice, advice on spares supply, with special insurance schemes for the cars, and a full calendar of social and competition events.

The monthly magazine is titled *The Jaguar Driver*, and goes to every member. Following the early-1980s strife, membership dropped from its peak, and in the early 1990s it stood at around 12,000. The JDC is independent of the factory itself, and there is a permanently staffed office at:

Jaguar Drivers' Club Ltd
Jaguar House
18 Stuart Street
Luton
Bedfordshire LU1 2SL
England
Tel: (0582) 419332

### Jaguar Enthusiasts' Club Ltd

This club was founded in 1984 as a break-away organization from the Jaguar Drivers' Club, to concentrate much more on the maintenance, restoration and remanufacturing side of Jaguar motoring. All the appropriate literature – workshop manuals, parts lists and a unique list of specialist suppliers – are available to members; there are special insurance schemes and the Club also offers specialist tools for hire, to aid in major rebuilding jobs. In the early 1990s the club was still growing, with more than 7,000 members, and there was an unstated ambition to overtake the size and importance of the older organization. The club's HQ is at:

JEC Ltd
Sherborne
Mead Street
Stoke Gifford
Bristol
Avon BS12 6PS
England
Tel: (0272) 698186

### Jaguar Car Club

This is much the youngest *and* smallest of the three organizations, for it was founded as recently as 1988, and has less than 500 members. The JCC was a splinter group from the JEC, and is heavily involved in Motorsport, with its own one-make series on British circuits. The club may be contacted through:

Badges are usually easy enough to find when a restoration is in progress, but grilles in good condition cost a great deal more.

*Every little detail – tail lamps, bumpers, overriders and decoration – is important when you are tackling a restoration. Parts for early cars are no longer available from the factory – but a great deal of re-manufacture has already taken place.*

Corrosion often starts close to door handles, not only because of scratching, but because of the stress applied to panel surfaces over the years.

*The art of a truly careful restoration is to make sure that panel shut lines are perfect, and that every little fitting is correctly aligned with its neighbour.*

The fuel filler cap on XJ-S cars often becomes strained, due to the hinges being bent, or goes rusty, or discoloured due to fuel spillage.

The V12 badge, and what looks like a tiny national motif, were personal affectations on this nicely presented XJ-S.

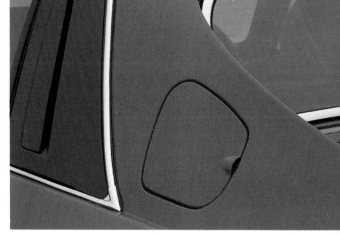

The most time-consuming part of any restoration is getting the detail right. Take the trouble to know what, and what material, should be used for every item of trim.

One of the first chrome parts to suffer on a Jaguar is the wing mirror, which may have been clipped by other mirrors during its life, or scratched by passers-by when parked.

Jaguar Car Club
Bluntington House
Chaddesley Corbett
Kidderminster
Worcestershire DY10 4NR
England
Tel: (0562) 83554

## Daimler & Lanchester Owners' Club

This very long-established club caters for owners of *all* Daimler and Lanchester cars, which means that there is a very strong emphasis on services for cars produced before the Jaguar take-over of 1960. Even so, a special register for the 'XJ-Daimler' cars is now in existence. In almost every way this duplicates the services and fellowship provided by the other Jaguar clubs. The HQ is at:

Daimler & Lanchester Owners' Club
Daimler/Lanchester House
Church Street
Gaminglay
Sandy
Bedfordshire
England
Tel: (0873) 890737

## OVERSEAS

The British Jaguar clubs have links with other clubs, which have been set up in many overseas countries, and in all cases *except* North America it is advisable to contact a British club for further information.

In North America, however, there is a blanket organization – the Jaguar Clubs of North America – which has close links with Jaguar's own North American organization, and is centred in New Jersey, not far from New York. All further details are available by contacting the HQ, at:

Jaguar Clubs of North America
555 MacArthur Boulevard
Mahwah
New Jersey
USA

# 12 What Next?

By the late 1980s, Jaguar's future was once again looking uncertain. In the years after privatization car sales had boomed, particularly to export territories like North America, and profits surged ahead. Then, by 1988, trends reversed, profits fell, and the company began to look vulnerable. This was when financial analysts pounced, crunched their numbers, and gave it as their opinion that Jaguar could not possibly stay independent for ever. There were two problems; one was that, although the company was still making money, it did not look as if it could ever generate enough funds, by cash flow, to invest in important new models, and the other was that the British government held the vital 'Golden Share' in the company's financial structure, which meant that no other company could take more than 15 per cent of Jaguar's capital without its blessing.

Jaguar's engineering capability, in any case, was woefully limited compared with that of its major rivals at Mercedes-Benz and BMW. Sir John Egan, a businessman to his boots, realized this when the XJ40 was being finalized, so while he concentrated on keeping the current cars' image as high as possible, he also decided to develop a new technical centre. It was Jaguar's great good fortune that Peugeot-Talbot, having decided to concentrate all its technical resources in France, left behind a good British technical centre at Whitley, on the southern outskirts of Coventry. Jaguar made haste to buy the plant, spent two years modernizing it to its own requirements, and finally moved in during 1987.

---

### The Whitley Technical Centre

Like many other industrial buildings in Coventry, the Whitley factory was originally built with military production in mind, the area having been used as a grass airfield since 1918. Positioned to the south-east of Coventry's city centre, it was well clear of any housing in the area (and therefore very secure), was used for aircraft testing in the 1920s, and later (after the first major buildings were erected) for aircraft manufacture by the Armstrong-Whitworth business.

The Armstrong-Whitworth Whitley bomber was assembled on the site in the 1930s and 1940s, but from 1948 onwards the factory worked exclusively on guided weapons development. By the 1960s, however, the site was empty, so in 1969 Armstrong-Whitworth sold it off to the Rootes Group (which had recently been taken over by Chrysler of the USA). Rootes-Chrysler redeveloped it as a technical centre, and by 1970 had moved all its design, engineering, product planning and styling functions on to this site.

Rootes became Chrysler UK in 1970, but after Chrysler sold out to Peugeot in 1978, it became Peugeot-Talbot Ltd. Peugeot then ran down its new-car design facilities in the UK, finally pulling out of Whitley and selling it to Jaguar in 1985.

Jaguar then completely re-equipped the 155-acre facility, erected several new buildings, made space for 1,000 engineers and their high-tech equipment, and admitted to spending £55 million on the job. The technical staff began moving in during 1987 and 1988. It was officially opened in May 1988.

Sixteen years after its initial launch, the XJ-S was face-lifted. Jaguar's new owners, Ford, saw that the car still had an excellent image, and that a freshen-up, to prolong its life, was justified.

1991-model XJ-S Coupés had different tail lamps, and many subtly reshaped panels, but the basic chassis and running gear was not changed. The optional six-cylinder engine was enlarged to 4.0 litres.

*From the side view, the major change for the 1991-model XJ-S was the reshaping of the rear quarter window, and the trim panelling surrounding it.*

*The most popular derivative of the 1991 XJ-Ss was the Convertible, which sold especially well in the USA.*

## Mergers and Manoeuvres

When it was formed, the Swallow Sidecar Company was a private concern, where William Lyons and Bill Walmsley were partners, as they also were in Swallow Coachbuilding and in SS Cars Ltd. From 1934, when Walmsley resigned, William Lyons became the major shareholder in SS Cars Ltd, and there he stayed until 1966.

SS Cars Ltd became Jaguar Cars Ltd in 1945, but it was not until the 1960s that Sir William began a series of major acquisitions. In May 1960, the BSA Group (which owned the Daimler company) agreed to sell Daimler-Lanchester to Jaguar. Sir William was not only interested in acquiring famous names, but the space in the Coventry factories. Before long he was to re-equip those factories to build engines and gearboxes for his cars, while the Daimler badge eventually appeared on cars like the Sovereign.

Sir William followed this up by acquiring the truck manufacturer Guy Motors (of Wolverhampton) in 1961, buying Coventry-Climax in 1963, and ending up by purchasing the Henry Meadows (engine manufacturing business) of Wolverhampton in 1964.

Then it was the turn of Jaguar to be progressively absorbed. In 1966, Sir William and his family still owned 260,000 of the 480,000 voting shares in the company, but to underpin the company's future, in July 1966 he agreed to sell out to the British Motor Corporation (BMC), which created British Motor Holdings.

BMH only had an eighteen-month existence, for in January 1968 it was submerged in the unwieldy colossus of British Leyland. Jaguar came close to losing its identity, and for a time operated as just one component of Jaguar-Rover-Triumph Ltd. Made autonomous again in 1978 by Sir Michael Edwardes' new regime it struck an all-time low in 1980, then recovered steadily.

Jaguar Cars was privatized in 1984, but could not stay independent indefinitely, and concluded an agreed take-over by Ford in 1989.

## RUMOURS

In the meantime, and having digested the launch of the new-generation XJ6, the British motoring press began spreading exciting rumours about Jaguar's future plans; Jaguar would launch a new V12-engined version of the XJ40 in 1988; Jaguar would soon produce a new two-seater sports car as a spiritual replacement for the old E-type; Jaguar's next major objective was to design a new smaller Jaguar, ostensibly to be a BMW 5-Series competitor, to be ready in the early 1990s . . .

By 1989, the rumour-mongers were still hard at work, but to reflect Jaguar's cash-flow problems their 'leaked' timetable had slipped considerably. The first of the new cars to come along – in 1992, it was said – would be the longer-wheelbase XJ40 body-shell, complete with its modernized V12 engine. Next up would be the XJ41/XJ42 projects, a pair of 2+2 sports coupé/cabriolet cars given the unofficial title of 'F-type', featuring four-wheel drive, and due to undercut (and, therefore, to undermine) the XJ-S range. Then, it was said, a new medium-sized Jaguar, coded 'XJ80', would follow, but probably not until 1996 . . .

Jaguar managers, who wanted to do *all* those things (and, incidentally, wondered how some of their long-term projects had leaked out), knew that they would probably be wishful thinking until the engineering capability was enlarged, and after new capital could be found, and said as little as possible.

## FORD OR GM?

Sir John Egan, for his part, had started looking around for a possible partner in March 1988, and by mid-1989 (when the company's shares were at a peak on the stock market) the City of London was openly talking about a future take-over battle.

In the autumn of 1989 events moved swiftly. In September, Jaguar announced a profits collapse for the first half of the year – from £22.5 million in 1988 to a mere £1.4 million in 1989 – which City analysts declared to be 'appalling'. Almost at the same time, Ford announced its intention to buy a stake in Jaguar, the British Prime Minister endorsed this as good for Jaguar – and it became clear that Ford's major rival, General Motors, was also shaping up with a financial proposal.

The two American giants, it seemed, were adopting entirely different approaches. GM wanted to take a minority shareholding, to offer financial seed-corn for the future, and to offer a gradual approach towards integration, rather as they had started to do with Lotus.

Ford, on the other hand, wanted to take financial control, to provide its own top managers and its own systems, and to turn Jaguar round without erasing its character. A Ford spokesman commented that 'We see ourselves as natural partners. We could offer Jaguar a great deal of support . . .'

When Ford announced its intention to buy Jaguar stock, the company's total worth in shares rose to more than £1 billion, but without actually spelling it out Ford discreetly made it clear that they had enough money to buy up *every* share when the time came.

Jaguar, at least in public, did not welcome this in-fighting, for as spokesman David Boole commented: 'We do not welcome Ford's approach. We want to remain independent and we feel that's the best way to serve the interests of our shareholders and customers . . .'

By early October 1989, it was clear that Sir John Egan's team had already been talking to General Motors for some time – months, indeed, before Ford declared an interest – but only on the basis of GM taking a minority shareholding. For Jaguar-watchers, if not for the company's managers themselves, it was a fascinating situation.

This was the crucial month. No sooner had Ford announced its intention to bid (which caused Jaguar's shares to rocket to £7.40 each) than GM let it be known that it was 'close to a deal' – but that this would mean taking only 15 per cent of the capital at first, and a total of 30 per cent after the 'Golden Share' time-limit expired at the end of 1990.

But GM, proving that old habits die hard, moved too slowly (it was not called a 'dinosaur' for nothing . . . ) and Ford, exasperated by the delays, decided to launch a full-scale bid. Having already bought up 12 per cent of the company for £140 million, it told the Government of its intention.

On Tuesday, 31 October, Government minister Nicholas Ridley stated that the 'Golden Share' restriction would be removed forthwith, and just one day later Ford offered £8.50 a share for Jaguar's entire capital. In City terms this was a 'knock-out' bid, for General Motors immediately withdrew from negotiations, leaving Ford's bid to become a formality. As Sir John later admitted, on the day he left Jaguar in June 1990:

The deal we were designing with General Motors was one we would have been proud to put in front of our shareholders. But we felt that if and when it arrived, the Ford one for 100 per cent control was going to be better. We had come to the conclusion that, on balance, the better owner was going to be Ford.

That was the conclusion we were coming to. That only with complete ownership would there be a full and proper technology transfer. We asked GM if they were interested in 100 per cent control, and gave them two weeks notice that a full bid from Ford was coming.

So why did GM not react?

Because they decided they wouldn't pay £1.6 billion. By then Ford was looking the better owner. They were more positive about the

*XJ-Ss continued to have a choice of six-
cylinder or V12 engines in the early 1990s.
This V12-engined example also had
Convertible coachwork.*

*Wood veneer, chrome, leather, and sleek detailing – the treatment of door panelling in the 1991 XJ-S.*

*Naturally the standard seat material for the 1990s-style XJ-S was leather.*

*For the 1991 cars, the fascia and instrument layout of XJ-Ss was once again revised.*

*The face-lifted XJ-S, identified here by reshaped tail lights and different badges, was unveiled in 1991 under Ford control.*

whole thing. From top to bottom they wanted to own Jaguar. There were conflicting messages coming out of General Motors. Some people did and some people didn't want Jaguar. Our discussions with GM didn't always go well . . .

Ford's take-over bid was agreed and formalized in a matter of days, and by the end of 1989 Jaguar had become a wholly owned subsidiary of Ford. Sir John Egan was not expected to stay on for long after the changeover was complete.

## THE FUTURE

In the next few months, Ford's influence on Jaguar was surprisingly limited. Ford of Europe's Vice-President of Business Strategy, John Grant, became Jaguar's Deputy Chairman in January 1990, while Lindsay Halstead (Ford of Europe's Chairman) also became a director. Then, at the end of March 1990, Sir John announced his intention to step down from the chair, his place to be taken by Ford's formidably talented Bill Hayden, who was then Vice-Chairman of Ford of Europe. Later in the year, Ford's Clive Ennos joined Jaguar as Director of Product Engineering, while at the same time Jim Randle moved sideways to look after the important job of concept engineering of future models.

In 1990 and 1991, Ford moved decisively to underpin Jaguar's future. Even before Sir John Egan stepped down, the XJ41 project was cancelled. This design, it seemed, was already long in the tooth, weighed a lot too much, and would have cost a great deal more than Jaguar had hoped. In the meantime, Bill Hayden had swept through the various Jaguar factories, had been staggered by the way that Jaguar technology was seen to lag behind that of the Ford factories, and had decided on a complete modernization.

---

**Bill Hayden (born 1929)**

When Sir John Egan left Jaguar at the end of June 1990, Bill Hayden, a Ford man to his toe-caps, took over as Chairman; he had become Jaguar's Chief Executive in March 1990. This was the clearest possible signal that Ford meant business, that Jaguar was going to be modernized, and that it was to be done in the most efficient way possible.

Bill Hayden, a Londoner born and bred, already had nearly forty years of Ford experience, had been appointed Ford's Vice-President of manufacturing in 1984, and was ready to take early retirement when the Jaguar appointment came along.

Hayden had always been a manufacturing man. His first, and immediate, assessment of Jaguar's business was that its build methods were well out of date, and that a shake-up was badly needed. He also confirmed that Ford had cancelled the F-type (XJ41 and XJ42) project, because it '. . . failed to meet one single performance objective: it was grossly, hundreds of pounds, overweight.'

In spite of undergoing open-heart surgery within months of his appointment, he was soon back at work, pressed ahead with a new business plan for the 1990s, and showed every sign of steering Jaguar through several more years of exciting rejuvenation.

---

Ford demanded, and got, a completely revised business plan for the Jaguar enterprise, and by 1991 it seemed to be clear that the four major plants – Telford for body stampings, Castle Bromwich for bodyshell manufacture, Radford for engine and component manufacture, and Browns Lane for car assembly – would be retained. Hayden and John Grant, it seemed, were convinced that Jaguar could be built up to produce 150,000 cars a year in these existing plants *if* the appropriate sort of high-technology equipment was drafted in. On that basis, Hayden admitted, Jaguar *is* interested in developing a new mid-sized car, and wants it to be 'a new type of Mk II'.

*Although the soft-top and hood frame didn't fold completely flat on the 1991 XJ-S, there was no rearward visibility problem.*

Where does this leave the XJ6 and XJ-S ranges in the early and mid-1990s? Quite literally, as the backbone of the company for some years to come. Millions of fine words don't alter the fact that completely new Jaguar models are still years away, and that Jaguar is relying on more and yet more versions of the existing six-cylinder and V12-engined cars to fill its production lines in the near future. Both the cars have years of refinement, development, and improvement ahead of them, but I think we can be sure of one thing – that Ford engines will never be used!

*Not even on the face-lifted XJ-S could the convertible soft-top be folded completely flat.*

# Index

Aston Martin, 34, 91
Aston Martin Tickford Ltd, 91
Austin Seven Swallow, 9, 10
*Autocar*, 13, 34, 35, 37, 42, 47, 83, 112, 146, 161, 162, 163, 166, 167, 170, 171, 172
*Autocar & Motor*, 54, 153, 159, 170, 171, 172
*Autosport*, 162, 163, 167, 170, 171

BAA, 130
*Back from the Brink*, 110
Baily, Claude, 12, 15, 39
Barber, John, 101, 116
Beasley, Mike, 144, 146
Bell, Derek, 116, 117, 119, 122
Bertone, 126, 127
Birkin, Sir Henry ('Tim'), 62
Black, Sir John, 10
BMC, 10, 31, 182
BMW, 34, 54, 111, 131
Bolster, John, 162, 163, 166
Boole, David, 147, 183
British Leyland (BL), 10, 34, 35, 38, 55, 62, 63, 70, 73, 83, 99, 101, 102, 103, 105, 106, 107, 110, 112, 113, 114, 119, 122, 126, 127, 130, 141, 144, 145, 147, 182
British Motor Holding (BMH), 34, 99, 102, 182
Broad, Ralph, 114
Broadspeed, 52, 114, 115, 116, 117, 118, 119
Brundle, Martin, 123
BSA, 17, 182

Cadillac, 40, 43, 44, 52
*Climax in Coventry*, 37
Cosworth, 123
Coventry-Climax Ltd, 38, 39, 40, 42, 182
Craft, Chris, 117
Crisp, Trevor, 131, 137
Crossley Motors, 9

Daimler company, 17, 182
Daimler models: 1898 model, 20
   3.6 litre/4.0 litre (XJ40 family), 150, 153, 173
   Double Six: Coupé, 52, 70, 173
     XJ12 type and Vanden Plas, 20, 40, 41, 46, 47, 49, 53, 163, 173
   Sovereign, 1966–1969, 18, 23, 35
     XJ6 type and Coupé, 35, 36, 40, 64, 69, 70, 112, 173

SP250, 18
V8-250 (2½ litre), 17, 18, 23, 35
Davis, S. C. H. ('Sammy'), 13
Dewis, Norman, 30, 31
Donnelly, Bill, 154, 155, 158, 160

*Economist, The*, 107
Edwardes, Sir Michael, 107, 110, 130, 182
Egan, Sir John, 83, 84, 110, 111, 112, 113, 122, 130, 131, 146, 154, 179, 182, 183, 187
England, 'Lofty', 46, 62, 63, 71, 99, 101, 102, 114, 126
Ennos, Clive, 131, 187
European Touring Car Championship (ETC), 115, 119, 123, 125

Ferrari, 34, 38
Fisher & Ludlow, 144
Fitzpatrick, John, 119
Ford, 67, 130, 131, 141, 150, 152, 160, 180, 182, 183, 186

Gardner, Fred, 22, 27
Gaydon proving grounds, 141
General Motors (GM), 130, 141, 182, 183, 187
Giugiaro, Giorgetto, 126
GKN, 97, 144
Grant, John, 187
Greville-Smith, Chris, 102
Group 44, 115, 122
*Guinness Book of Records*, 126
Guy Motors, 182

Hahne, Armin, 125
Halstead, Lindsay, 187
Hassan, Walter, 12, 15, 37, 38, 39, 40, 42, 107
Hayden, Bill, 130, 159, 187
Henlys Ltd, 10
Heyer, Hans, 123
Heynes, William, 12, 15, 16, 21, 30, 39, 71, 79, 107
Hobbs, David, 116
Holtum, Colin, 98, 102
Horrocks, Ray, 110, 113, 130

Ital Design, 126, 127

Jaguar Cars Ltd, 13 *et seq.*, 14, 182
Jaguar Daimler Heritage Collection, 113

*Jaguar Driver, The*, 175
Jaguar (and SS) factories:
    Blackpool, 9, 10
    Browns Lane, 14, 15, 17, 35, 45, 53, 63, 82, 83,
        110, 112, 113, 114, 126, 127, 137, 144, 145,
        146, 156, 157, 187
    Castle Bromwich, 83, 91, 96, 111, 112, 113, 144,
        145, 146, 187
    Foleshill, 10, 13, 14, 15, 22
    Radford, 17, 38, 41, 46, 63, 113, 135, 144, 145,
        146, 187
    Telford, 144, 187
    Vanden Plas, North London, 47
    Whitley, 98, 130, 146, 179
Jaguar (and SS) models: 2.4/3.4/3.8 family, 11, 12,
    14, 16, 18, 29
    420 family, 17, 18, 19, 23, 24, 25, 30, 31, 32, 35
    C-Type, 12, 15, 62, 79, 107
    D-Type, 12, 14, 15, 30, 38, 62, 71, 79, 82
    E-Type, 12, 16, 17, 19, 20, 25, 30, 35, 38, 39, 42,
        43, 71, 72, 76, 79, 82, 83, 106, 113, 114, 135,
        146, 174, 182
    Mk II/240/340 family, 17, 18, 19, 23, 24, 25, 34,
        35, 174, 187
    Mk V, 14, 135
    Mk VII family, 11, 12, 14, 15, 16, 17, 22, 23, 144
    Mk X and 420G, 17, 22, 23, 24, 30, 32, 35, 42
    SS1, 11, 22
    SS2, 11
    SS100, 12, 13
    SS Jaguar, 11, 12, 13
    S-Type family, 17, 18, 19, 23, 24, 30, 34, 35
    XJ6 and XJ6C, 9, 10, 12, 15, 16, 17, 19, 20,
        21–35, 36, 37, 39, 40, 42, 43, 44, 51, 52, 53,
        55, 56, 59, 60, 61, 63, 65, 66, 67, 68, 70, 71,
        76, 78, 96, 99, 101, 102, 103, 105, 106, 107,
        108, 109, 111, 112, 113, 114, 126, 130, 131,
        134, 135, 139, 145, 147, 150, 151, 161, 162,
        163, 173
    XJ6 family (coded XJ40), 10, 21, 45, 53, 54, 99,
        102, 103, 111, 113, 126, 127, 130, 131, 132,
        133, 134, 135, 136, 137, 139, 140, 141, 142,
        143, 144, 146, 147, 148, 150, 151, 152, 153,
        159, 173, 182, 188
    XJ12 and XJ12L, 34, 37, 39, 40, 42, 43, 44, 45,
        46, 50, 51, 52, 78, 82, 86, 114, 115, 126, 127,
        145, 146, 150, 152, 153, 163, 166, 173
    XJ12C and XJ5.3C, 51, 52, 67, 70, 115, 116, 117,
        118, 119, 122, 123, 166, 173
    XJ13, 38, 39
    XJ41/XJ42 ('F-Type'), 182, 187
    XJ80, 182
    XJ220, 131, 159, 160
    XJR-5 race car, 115
    XJR-7 race car, 115
    XJR-8 race car, 115

    XJR-12 race car, 115, 160
    XJR-15 race car, 159
    XJR-S Coupé, 155, 156, 157, 158, 159, 160, 161
    XJ-S (coded XJ27) and XJ-S HE, 10, 19, 55, 63,
        71–98, 101, 102, 107, 115, 116, 122, 123, 124,
        125, 126, 135, 136, 137, 141, 145, 146, 148,
        149, 154, 155, 156, 164, 165, 166, 170, 173,
        177, 180, 181, 182, 185, 186, 188
    XJ-SC and Convertible, 90–3, 95, 96, 97, 98, 144,
        145, 146, 149, 152, 171, 173, 181, 184
    XK120/XK140/XK150, 12, 14, 15, 17, 23, 135,
        174
Jaguar-Rover-Triumph Ltd, 107, 110, 182
Jaguarsport Ltd, 154–60
Jones, Tom, 30

Karmann, 96
Knight, Bob, 12, 30, 31, 43, 63, 71, 82, 101, 106,
    107, 130, 131

Lee, Leonard, 38
Le Mans 24 Hour race, 15, 30, 39, 62, 115, 156, 160
Leyland Cars, 102, 105, 115, 119
Leyland Motors (*see* British Leyland)
Lord Wakefield Gold Medal Paper, 35
Lyons, Sir William, 9, 10, 11, 12, 13, 14, 15, 16, 19,
    20, 21, 22, 23, 24, 25, 31, 34, 35, 38, 52, 60,
    62, 68, 71, 111, 113, 126, 127, 182

Maserati, 38
May, Michael, 46, 52, 84, 139
McCarthy, Mike, 171
Meadows, Henry, Ltd, 182
Mercedes-Benz, 30, 34, 38, 40, 44, 54, 83, 98, 111,
    114, 131
MIRA, 39
*Modern Motoring*, 35
Morris Cowley Swallow, 9
*Motor*, 83
Mundy, Harry, 37, 38, 39, 40, 42, 43, 63, 131, 137

National Enterprise Board, 107, 110
*New Avengers, The*, 70
Nicholson, Chuck, 123

Orr-Ewing, Hamish, 113, 130

Park, Alex, 114, 126
Park Sheet Metal Co., 91
PBB, 97
Percy, Winston, 123, 125
Pininfarina, 99, 101, 102, 105, 106, 126, 127
Porsche, 90
Pressed Steel Co. Ltd (Pressed Steel Fisher), 16, 25,
    30, 31, 34, 62, 73, 83, 130, 144, 145

Randle, Jim, 98, 130, 131, 134, 135, 141, 144, 187
Ricardo, 52
Ridley, Nicholas, 183
Robinson, Geoffrey, 62, 63, 99, 102, 114, 126
Rolls-Royce, 30, 40, 44, 52
Rouse, Andy, 116, 119, 122
Rover, 150
Ryder, Lord, 126
Ryder Report, 102, 105, 107, 114

Sayer, Malcolm, 10, 71, 78, 79, 99
Schenken, Tim, 119, 122
SS Cars Ltd, 10, 11, 12, 39, 182
Standard Motor Co. Ltd, 10
Standard Sixteen, 10
Stokes, Lord (earlier Sir Donald), 36, 62, 99, 101, 102, 114, 126
Straight, Whitney, 62
Swallow Coachbuilding, 9, 10, 22, 182

Thompson, Steve, 116, 117
Thorpe, Doug, 102
Tickford, 91
Tourist Trophy race, 116, 117, 118, 122
Tullius, Bob, 115, 122
TWR, 91, 93, 115, 122, 123, 124, 130, 154, 155, 156, 159, 160

Venture Pressings, 97

Waeland, Derek, 146
Walkinshaw, Tom, 115, 122, 123, 125, 130, 154, 159, 160
Walmsley, William, 9, 10, 182
Weslake, Harry, 12, 52
Whittaker, Derek, 115, 126
Whyte, Andrew, 20
Wilson, Berry, 110
Winterbottom, Oliver, 102
World Sports Car Championship, 123